Energy Storage Applications in Power Systems

Edited by Zongjie Wang and Abdollah Younesi

Published in London, United Kingdom

Energy Storage Applications in Power Systems
http://dx.doi.org/10.5772/intechopen.104058
Edited by Zongjie Wang and Abdollah Younesi

Contributors
Chukwuemeka Emmanuel Okafor, Komla Agbenyo Folly, Ya LV, Seyyed Mohammad Sadegh Ghiasi, Narges S. Ghiasi, Ahmad Alaa Mahfouz, Kiswendsida Elias Ouedraogo, Pinar Oguz Ekim, Erhan Demirok, David R. Walwyn

First published in London, United Kingdom, 2023 by IntechOpen
IntechOpen is the global imprint of INTECHOPEN LIMITED, registered in England and Wales, registration number: 11086078, 5 Princes Gate Court, London, SW7 2QJ, United Kingdom

British Library Cataloguing-in-Publication Data
A catalogue record for this book is available from the British Library

Additional hard and PDF copies can be obtained from orders@intechopen.com

Energy Storage Applications in Power Systems
Edited by Zongjie Wang and Abdollah Younesi
p. cm.
Print ISBN 978-1-83768-679-7
Online ISBN 978-1-83768-680-3
eBook (PDF) ISBN 978-1-83768-681-0

Meet the editors

Dr. Zongjie Wang's research interests focus on modern power systems, grid resilience, and renewable energy integration through leveraging data analytics, non-linear optimization, and coordinated transmission and distribution (T&D) systems. Her projects include the development of new algorithms for OPF in power systems with high penetration of renewable energy sources, bi-level optimization between T&D systems, and so on. As an invited speaker, Dr. Wang presented her research at the headquarters of the US Federal Energy Regulatory Commission's technical meeting in Washington, DC, in 2019 and 2023. Dr. Wang is a member of the Institute of Electrical and Electronics Engineers (IEEE) and the Power & Energy Society (PES). She has been a member of many IEEE committees such as the PEEC committee. She has served as a guest editor for many journals, including *Frontiers in Energy Research*. Dr. Wang has collaborated with many power system researchers in industries and institutions such as ISOs, the Massachusetts Institute of Technology (MIT), and Cornell, among others.

Abdollah Younesi received his BSc, MSc, and Ph.D. in Electrical Engineering from the Faculty of Technical Engineering, Mohaghegh Ardabili University, Ardabil, Iran in 2012, 2015, and 2021, respectively. He was a senior research associate in the School of Computing Science, University of East Anglia, Norwich, UK in 2022. He is a Ph.D. researcher in the Department of Electrical Engineering, University of Connecticut, USA. Throughout his academic career, Dr. Younesi has been selected as a Talented Student of the University several times. In addition, he was twice chosen as the top talent of the National Elite Foundation in 2018 and 2019. His areas of interest are power system decarbonization, power system non-linear optimization and simulation, application of Reinforcement Learning (RL) to power system control, time series analysis and forecasting, resilient operation of power systems, smart grid technology, renewable and storage energy, dispatch control hierarchy, data analytics, engineering management, active distribution systems, microgrids, and digital image processing.

Contents

Preface

In this age of technology-driven revolution, our reliance on energy systems has never been more pronounced. Power systems, once deemed adequate with fossil-fuel-based energy sources, are now undergoing a significant transformation to accommodate renewable energy sources and advanced energy storage systems. This edited volume, *Energy Storage Applications in Power Systems*, brings together pioneering research and innovative ideas in energy storage and its diverse applications.

Kicking off the volume, Chapter 1, "Mechanical Energy Storage Systems and Their Applications in Power Systems", provides an overarching view of mechanical energy storage technologies and their potential to enhance grid stability. Chapter 2, "A Review on Superconducting Magnetic Energy Storage System Applications", shifts focus to the specialized field of Superconducting Magnetic Energy Storage (SMES), discussing its unique characteristics and potential benefits to power quality and system stability. Continuing the examination of specific technologies, Chapter 3, "Battery-Assisted PV-Pumping System", elucidates the intersection of photovoltaic (PV) systems and energy storage in pumping applications. It explores how battery energy storage can enhance the reliability and efficiency of these systems. Transitioning from technology-focused to application-based discussion, Chapter 4, "Energy Storage System (ESS) in Residential Applications", showcases the potential of residential energy storage in transforming power systems and fostering a community-centric energy ecosystem. Building upon this, Chapter 5, "Balancing Renewable Energy Capacity, Time of Use Tariffs and Energy Storage in Energy Systems", offers a wider perspective on harmonizing renewable energy, time-of-use tariffs, and energy storage to optimize energy utilization and reduce costs. Rounding out the volume, Chapter 6, "Prospection of Neighborhood Megawatthours Scale Closed Loop Pumped Hydro Storage Potential", takes a deep dive into the potential of neighborhood-level pumped hydro storage. This chapter examines how such systems can counter fluctuations in renewable energy and contribute to grid stability.

My co-editor Dr. Abdollah and I have joined forces to ensure that this volume not only provides state-of-the-art insights but also fuels further inquiry and innovation in this pivotal field. The creation of this volume has been a rigorous but rewarding process, demanding meticulous attention to detail and a relentless pursuit of quality and relevance. Our gratitude extends to the dedicated contributors of this volume, a diverse ensemble of researchers and practitioners whose shared passion for innovation in power systems shines through in each chapter. Their hard work and dedication have made this volume a repository of knowledge where rigorous science meets practical application. We are equally thankful to the team at IntechOpen. Their support, professionalism, and commitment to academic publishing have been invaluable throughout the editing process, from the initial concept to the final product.

The goal of this volume is more than just the dissemination of scientific knowledge. It is also to serve as a beacon, guiding researchers, academics, industry practitioners, and policymakers through the rapidly evolving landscape of power systems. By delving into the intricacies of energy storage applications, we aim to spark further inquiry, encourage debate, and inspire innovative solutions.

The path toward a sustainable energy future is fraught with challenges, but each chapter in this volume testifies to the leaps of progress we have made and the potential for further advancement. As we turn the page on a new era of energy storage applications in power systems, this book serves as both a testament to our journey thus far and a roadmap for the terrain yet to be explored.

Zongjie Wang
Professor,
Department of Electrical and Computer Engineering,
University of Connecticut,
Storrs, CT, USA

Abdollah Younesi
Department of Electrical and Computer Engineering,
University of Connecticut,
Storrs, CT, USA

Chapter 1

Mechanical Energy Storage Systems and Their Applications in Power Systems

Chukwuemeka Emmanuel Okafor and Komla Agbenyo Folly

Abstract

The negative environmental impacts of conventional power generation have resulted in increased interest in the use of renewable energy sources to produce electricity. However, the main problem associated with these non-conventional sources of energy generation (wind and solar photovoltaic) is that they are highly intermittent and thereby result in very high fluctuations in power generated. Hence, mechanical energy storage systems can be deployed as a solution to this problem by ensuring that electrical energy is stored during times of high generation and supplied in time of high demand. This work presents a thorough study of mechanical energy storage systems. It examines the classification, development of output power equations, performance metrics, advantages and drawbacks of each of the mechanical energy storage types and their various applications in the grid networks. The key findings in this work are the strategies for the management of the high costs of these mechanical storage devices. These include deployment of hybrid energy storage technologies, multi-functional applications of mechanical energy storage systems through appropriate control methodologies and proper sizing strategies for cost effectiveness and increased penetrations of renewable energy sources in the power grid.

Keywords: mechanical energy storage, renewable energy, intermittent, performance measures, power systems

1. Introduction

Until now, the entire energy sector depends on fossil fuels for the generation of electricity, but more environmentally friendly options are advancing in the form of renewable energy sources. The transition from conventional (traditional) power plants to more environmentally friendly options will necessitate a need for more flexibility in the generation, transmission, and consumption of electricity. Energy storage systems (ESSs) can provide the flexibility that is needed for a robust high quality stable electrical system when technically integrated into the grid network. The following are some of the features of energy storage:

- Being able to store energy at the time of excess electricity production and making it available for use within a very short time window when the need arises.

- With the help of energy storage technologies, energy can be stored and made available at the very point where it is needed whether at the transmission, distribution, or consumption levels. This flexibility ensures the postponement of infrastructural upgrades in generation, transmission, and distribution networks.

- The consumption of energy varies with time. Sometimes the demand is high while at other times the demand is low. Energy storage devices can be deployed to meet the varying energy demands per time.

- Energy storage technologies such as pumped-hydroelectric storage (PHS), battery energy storage system (BESS), supercapacitors, etc. are flexible in providing multiple services to the grid. They can serve as loads during their charging process and therefore offer a service to the grid like voltage rise mitigation, while in their discharging mode, they can be controlled to provide peak-shaving service, frequency support or inertia support.

- Finally, energy storage technologies have the flexibility of participating in some special services to the grid for example the black start service. The system offering the black start service must possess the capacity to move from shutdown into operation without the aid from the grid. This type of service is required during periods of persistent blackouts. Energy storage technologies with high energy capacity like PHS, compressed air energy storage (CAES), and gravity energy storage (GES) can provide excellently the black start service to the grid.

There are six different categories of ESS, and these are: mechanical, thermal, chemical, electrochemical, electrical and hybrid system. Each category has unique characteristics in terms of life cycle, discharge time, discharge loss, energy density and power rating. All these characteristics account for their suitability for specific applications in the power system.

In mechanical energy storage system (MESS), there is a conversion of energy from mechanical to electrical form [1]. In times of low energy demands, electrical energy is taken from the grid and stored until the time of high demand when it is then converted back to electrical energy and transmitted back to the grid [2]. The flexibility in the conversion processes of MESSs accounts for their global applications [3]. MESSs are classified as pumped hydro storage (PHS), flywheel energy storage (FES), compressed air energy storage (CAES) and gravity energy storage systems (GES) according to [1, 4]. Some of the works already done on the applications of energy storage technologies on the grid power networks are summarized on **Table 1**.

Considering the works summarized in **Table 1**, the authors have done extensive research on energy storage integration to the grid network taking into accounts several aspects such as energy storage technology types, applications (both single and combined), limitations and challenges of energy storage systems, power electronic converters for energy storage interface. Simulation tools (software) for energy storage systems and storage system placement and sizing. However, to the best of our knowledge, nothing has been done on how to manage the high cost of energy storage system which according to [22] is the main reason hindering the massive deployment of energy storage system in the grid.

References	Contributions	Years of publication
[1, 3, 5–7]	The works present an in-depth review of energy storage technology types and their applications in the grid power networks.	2019, 2014, 2011, 2009, 2013
[8, 9]	The papers present the economic and reliability impacts of energy storage systems in power system networks.	2018, 2021
[10, 11]	The works discuss the application of energy storage systems in different levels of grid voltage. Besides, the conditions for integration of energy storage into the grid for proper compatibility with the operational codes and standards were emphasized.	2016, 2017
[12–14]	The authors explore the possible approaches of combining applications of energy storage systems. The technical requirements for the combination of applications were also discussed.	2004, 2003, 2019
[15, 16]	The works evaluate the challenges militating against the massive deployment of large-scale energy storage technologies for grid applications considering the various economic, legislative, and technical aspects.	2016, 2020
[17]	The work provides an in-depth review of the methodologies of storage sizing and placement on the grid networks. It covers several areas such as analytical approach, mathematical programming, exhaustive search, and heuristic methods.	2016
[18–20]	The papers perform a detailed analysis of power electronics converters used in interfacing energy storage systems with the grid network.	2016, 2017, 2020
[21]	The work discusses some of the software used in the simulation and analysis of energy storage systems and specific energy storage applications they are designed to implement.	2017
[22]	The work evaluates the impact of energy storage systems on the economic operation of distribution systems	2020

Table 1.
Summary of the works done on the applications of energy storage in the grid networks.

Therefore, the principal contribution of this work is bringing to light certain key approaches that should be addressed to manage the high costs of energy storage systems, thereby promoting their massive deployment into the grid networks.

The block diagram showing a simple classification of mechanical energy storage systems according to [23, 24] is given in **Figure 1**.

2. Pumped hydro-electric storage (PHS)

The PHS is a utility-scale energy storage technology that has been in implementation since 1890s [25]. It has a high commercial acceptance, and it is well-established. In PHS, large amount of water is delivered to an upper reservoir, during period of excess generation of electricity and it is converted back to electricity through the use generator and turbine during period of shortage of electricity. An illustration of pumped hydro-electric storage is shown in **Figure 2** while **Table 2** shows the performance measures.

It was reported in [26] that one of the earliest technologies for the storage of energy is PHS. The operating cost for energy units for PHS as compared to other energy storage systems has been reported to be the cheapest according to [18].

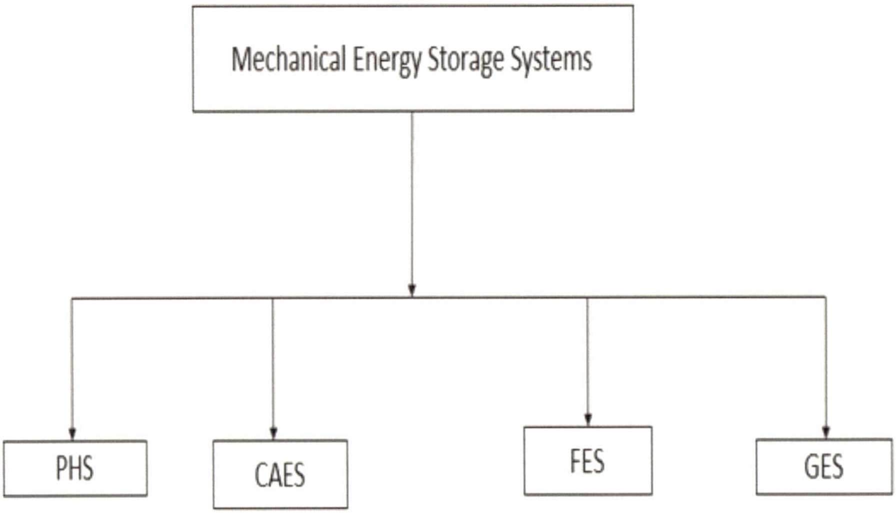

Figure 1.
Block diagram of mechanical energy storage systems.

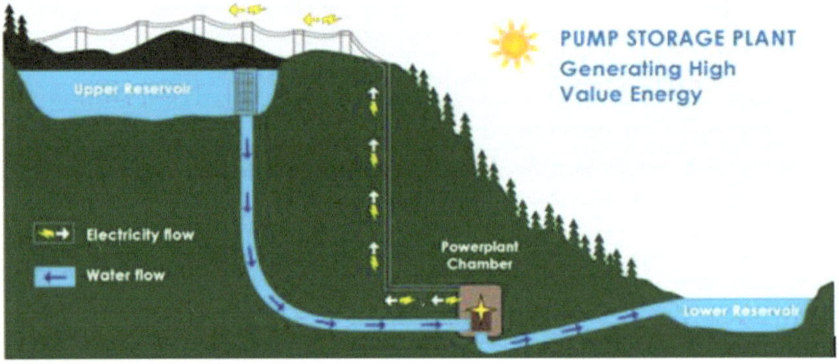

Figure 2.
An illustration of pumped hydroelectric storage [8].

In 2017, there were about 270 PHS stations globally producing 127 GW of electricity [27]. The United States and the European Union (EU) have 40 and 160 PHS stations respectively while the rest of the world has 70 PHS units [27].

2.1 Output energy equation of a PHS

Pumped-hydro-electric storage is generally used for energy-based applications because of its ability to deliver power for very long period in several hours [28]. It functions by utilizing the potential energy of water due to the force of gravity. When using the power from the grid during a period of low demand, water is pumped from lower reservoir to the upper reservoir. In the time of high demand of power, the water stored in the upper reservoir is released into the lower reservoir for the operation of the turbine and generator in order to inject power into the grid [24].

Efficiency (%)	70–85
Number of cycles (in cycles)	10,000–35,000
Expected life(years)	30–60
Specific energy (Wh/kg)	0.5–1.5
Specific power (W/kg)	N/A
Power capacity cost ($/kW)	600–2000
Energy capacity cost ($/kWh)	0–23
*BOP ($/kWh)	270–580
Power conversion system (PCS) ($/kW)	0–4.8
**O&M ($/kW-yr)	3–4.4
Maturity of the technology	Mature/ Implemented.

Adapted from [13]
**BOP: Balance of payment. **O&M: Operation and maintenance.*

Table 2.
Performance measures of PHS.

The energy stored, according to [29], depends on the volume of the water and the height of the waterfalls. This is described in Eq. (1).

$$E(phs) = \ell ghV \quad (1)$$

where $E(phs)$ is the stored energy in joules, ℓ is the density of water equivalent to 1000 kg/m^3, g is the acceleration due to gravity equivalent to 9.8 m/s^2, h is the height of the waterfalls (in meters) and V the volume of water stored in the upper reservoir in m^3.

2.2 Advantages and drawbacks of PHS

PHS has a high-power capacity ranging from some MW to about 3GW with a cycle efficiency of approximately 70–85 cycles and over 40 years lifetime [4, 6]. However, the drawbacks of a PHS lie in getting available sites to accommodate two large reservoirs and dams, the long-time involve in the preparation of the site, a high capital cost (in hundreds to thousands of millions of dollars) for construction, and environmental issues (removing trees and vegetation from a large amount of land) prior to the reservoir being flooded [30, 31]. The United States has an existing 23GW of PHS capacity installed [32].

3. Compressed air energy storage (CAES)

In this storage system, power from the grid (during period of low demand) is used to pump air into underground geological formation until the air is at high pressure [33]. During discharge, air at high pressure is drawn from the storage cavern and undergoes a heating and expansion process inside high- and low-pressure turbines, for conversion into kinetic energy and thereafter, transformed into electrical energy in a generator [29, 34]. One innovative system in the application of CAES is by combining it with a wave energy system according to [6]. The well turbine being an integral component of the wave energy system [35] uses the excess power produced by the renewable energy sources during the period of peak production. Through these methods, most of the renewable energy generated during the period of peak production would be used if not immediately. The excess power generated during peak production is used in storing fresh air in salt caverns [36]. The CAES turbine, using this fresh air can generate 3 times the output power for the same natural gas input [37]. With this innovation, a higher efficiency of the system is achieved.

An illustrative topology of a CAES is shown in **Figure 3** while the performance metrics are given in **Table 3**.

The current technological advancement on the improvement of the efficiency of CAES is focused on developing an advanced adiabatic CAES (AA-CAES) through which air is adiabatically compressed and pumped into an underground cavern. The success of this depends on the compressor and the expanded trains of the CAES [33].

In natural gas power plants with attached CAES, the air at a high pressure is mixed with natural gas for combustion [38]. The integration of CAES systems to an existing grid is relatively easy due to its similarity to a conventional gas combustion system [39].

CAES can be effectively utilized through the implementation of a hybrid energy storage system. A hybrid energy storage system involves the integration of different energy storage technologies for the implementation of several functions in the grid network [40]. To achieve this, the technical characteristics of the different storage technologies such as the power and energy capacities, response time and discharge time should be known. Besides, the application technical features should be analyzed for proper hybridization of different technologies. Examples of such hybridization include, CAES with flywheel examined in [40], CAES and supercapacitor energy storage and pumped hydro energy storage with CAES in [7]. CAES creates a potent energy reserve [41] and has three main components namely compressors, air storage reservoir and expanders [42].

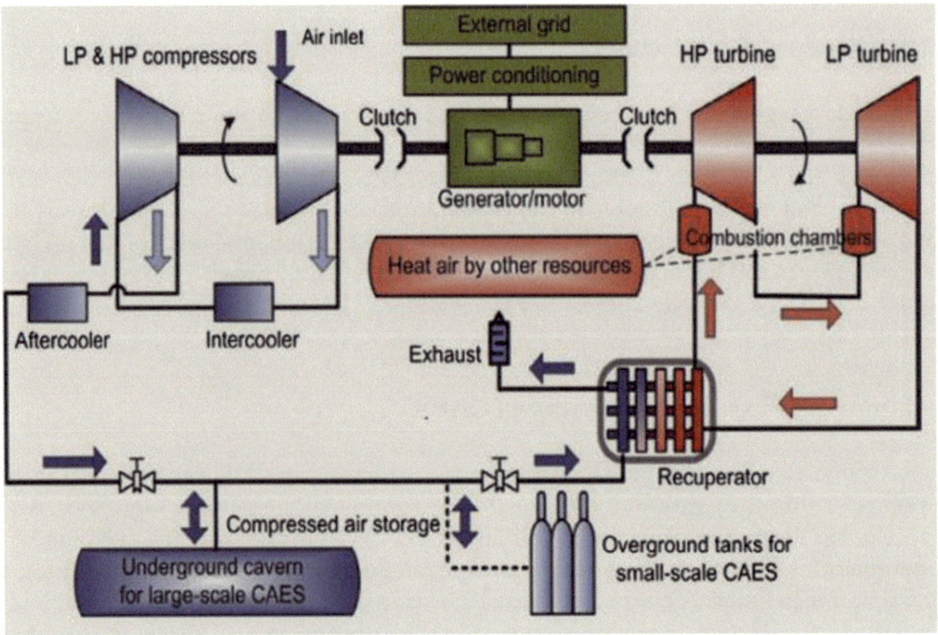

Figure 3.
An illustrative topology of a CAES [8].

Efficiency (%)	57–85
Cycle lifetime (cycles)	N/A
Expected lifetime (years)	20–40
Specific energy (Wh/kg)	6–30
Specific power (W/kg)	N/A
Power capacity cost ($/kW)	400–800
Energy capacity cost ($/kWh)	2–140
BOP ($/kWh)	270–580
*PCS ($/kW)	46–190
O&M ($/kW-yr)	1.6–29
Maturity	Commercial

Adapted from [13]
**PCS: Power conversion system.*

Table 3.
Performance measures of CAES.

3.1 Advantages and drawbacks of CAES

CAES has a large energy capacity, little geographical dependency, long lifespan, and low cost per kW [43]. The CAES system has a lifetime of about 40 years with an energy efficiency of 71% [29]. CAES owing to its installed capacity which is from 35 to 300 MW is deployed in the grid to support load leveling, voltage, and frequency control, etc., [44].

However, CAES suffers the same limitations as PHS, that is difficulty in locating a specific geographic place for installations. To overcome this limitation, high-pressure carbon fiber tank air storage is proposed for the implementation of distributed CAES system [12, 45, 46].

Also, CAES installations have relatively low efficiency [33]. The recent developments of hybrid CAES plants with offshore and onshore wind plants shows increased overall efficiency with reduced fluctuations in the power output.

3.2 Output energy equation of a CAES

CAES is an energy-based storage system that utilizes the principle of the gas turbine to produce electricity. Excess electricity (during periods of peak production) is used to compress and store air at very high pressure. When electricity is needed, the stored air at high pressure is deployed to drive a turbine to generate electricity [47, 48].

The power taken by each compressor (P_C) is given in Eq. (2) according to [49].

$$P_C = \left(\frac{K}{K-1}\right)\left(\frac{QRT_{in}}{\eta_c}\right)\left[\left(\beta^{(K-1)/K} - 1\right)\right] \tag{2}$$

where K is the adiabatic exponent, Q, the mass flow rate of air, R is the gas constant, T_{in} is the temperature inlet of the compressor, η_c is the efficiency of the compressor and β is the compressor ratio. The air temperature of the outlet (T_{out}) of each compressor stage could be expressed as in Eq. (3) according to [49].

$$\mathrm{T_{out}} = \mathrm{T_{in}} \left[\frac{\beta^{(t-1)/k} - 1}{\eta_c} + 1 \right] \tag{3}$$

Considering a compressor with many stages, n and charging time (t_c), the total energy taken by the compressor (W_c) can be evaluated using Eq. (4).

$$\mathrm{W_c} = \sum_{i=1}^{n} P_c t_c \eta_c \tag{4}$$

where η_c is the efficiency of the air compressor. Applying the adiabatic efficiency to compute the actual shaft power of the turbine, the power of each turbine (P_e) is given in Eq. (5) according Ref. [49].

$$\mathrm{P_e} = \frac{K}{K-1} (\mathrm{QRT}\eta_e) \left[1 - \left(EXR^{(K-1)/K} \right) \right. \tag{5}$$

where, Q is the mass flow rate of the turbine, η_e is the turbine efficiency, EXR is the expansion ratio of the turbine and T is the temperature of the turbine.

For a multi-stage turbine of m-number of stages, having a discharging time of t_e, the total electrical power output (W_e) is calculated using Eq. (6).

$$\mathrm{W_e} = {}_{\mathrm{g}} \sum_{t=1}^{m} P_e t_e \tag{6}$$

where, η_g is the efficiency of the generator.

3.3 Improving the efficiency of a CAES system

Energy storage efficiency is the principal factor militating against the development of CAES [49]. The energy efficiency of CAES depends on the energy efficiencies of all the units making up the CAES. These include the compression unit, air storage unit, heat regeneration unit and turbine generation unit. Thus, improving the performance of each of these units enhances the efficiency of CAES.

3.3.1 Compression subsystem

In order to have a compression subsystem with a high efficiency the exhaust temperature of the compression should be increased. This will result in an improvement in the heat storage temperature of the system and its storage efficiency.

3.3.2 Air storage subsystem

Underground salt cavern can be used to store large amount of air, and this will reduce the pressure fluctuations and consequently improves efficiency [49].

3.3.3 Regeneration subsystem

The efficiency of CAES is directly proportional to the temperature of the heat regeneration subsystem. As the temperature of the regeneration unit increases, the system efficiency also increases. In places where there is abundant supply of wind and solar energy, they can be adequately harnessed to provide heat for the regeneration subsystem, thereby increasing its temperature.

3.3.4 Turbine generator subsystem

The turbine generator unit is one of the major component parts of a CAES. It participates in the thermoelectric conversion in the energy-discharging process. The development of new and efficient air turbine will help in increasing the efficiency of CAES in general.

4. Flywheel energy storage (FES)

A flywheel energy storage (FES) is a rotating disk that can store or dissipate mechanical kinetic energy utilizing rotatory inertia [16]. An illustrative topology of an FES is shown in **Figure 4** and its performance metrics is given in **Table 4**.

In FES energy is stored in the angular momentum in a rotating mass [46]. Unlike PHS and CAES, FES is a power-based energy storage system [50]. It is deployed in applications that require short duration with short discharge time in the range of 1–100 s [37]. FES is used for voltage support [28], frequency support [37], fluctuation suppression, and provision of short-duration power quality. Globally, several FES systems are in use with an installed capacity of more than 940 MW [36].

4.1 Energy output and shaft factor equations of a flywheel

The rotating mass(disk) is mechanically connected to the shafts of the machines [29]. The energy stored by flywheels is given in Eq. (7) according to [48].

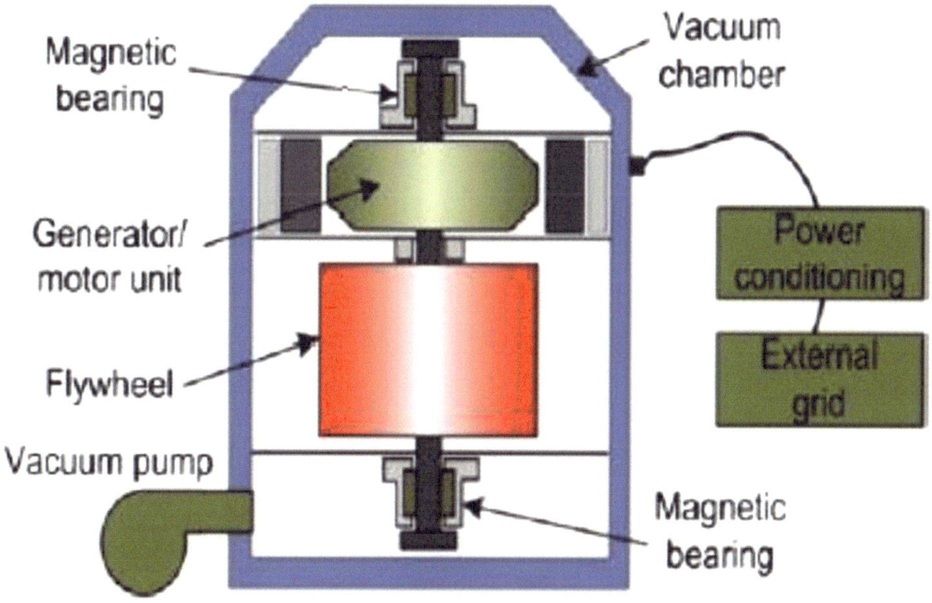

Figure 4.
An illustrative topology of a FES [8].

	Flywheel (low speed)	Flywheel (high speed)
Efficiency (%)	70–95	70–95
Cycle lifetime (cycles)	20,000–100,000	20,000–100,000
Expected lifetime (years)	15–20	15–20
Specific energy (Wh/kg)	10–30	10–30
Specific power (W/kg)	400–1500	400–1500
Power capacity cost ($/kW)	250–360	250–400
Energy capacity cost ($/kWh)	230–60,000	580–150,000
BOP ($/kWh)	110–600	110–600
PCS ($/kW)	0–120	0–1200
O&M ($/kW-yr)	6–22	6–22
Maturity	Developed	Developed

Adapted from [13].

Table 4.
Peformance measure of Flywheel energy storage.

$$E_{fes} = \frac{1}{2} J W^2 \tag{7}$$

where E_{fes}, (joules) is the energy stored by flywheel, J (in kgm^2) is the inertia of the rotating parts, W is the rotational speed in rad/s. The moment of inertia J is defined according to Eq. (8)

$$J = \int X^2 dM_x \tag{8}$$

where X is the distance from the axis of rotation, differential mass = dM_x. Considering a flywheel with radius r, having a mass M concentrated in the rim, the solution of the integral equation becomes Eq. (9).

$$J = \int X^2 dM_x = Mr^2 \tag{9}$$

The energy stored by the flywheel becomes $E_{fes} = \frac{1}{2}JW^2$, where W is the angular velocity. Thus Eq. (9) is transformed into Eq. (10).

$$E_{fes} = \frac{1}{2} Mr^2 W^2 \tag{10}$$

The energy density (E_{mass}) is given in Eq. (11).

$$E_{mass} = \frac{1}{2} r^2 W^2 \tag{11}$$

To obtain the volume energy density, (E_{volume}), we divide Eq. (10) by the volume V, therefore we have the volume density given as in Eq. (12)

$$\mathrm{E_{volume}} = \frac{E_{fes}}{V} = \frac{1}{2}\frac{M}{V}r^2W^2 \tag{12}$$

$$\text{But density } (\rho) = \frac{mass}{volume} \tag{13}$$

The density is given by Eq. (13)
Therefore, the volume energy density is now expressed as shown in Eq. (14).

$$\mathrm{E_{volume}} = \frac{1}{2}\rho r^2 w^2 \tag{14}$$

For the flywheel's rim, the tensile stress is given by σ, which is defined in Eq. (15) according to Ref. [36].

$$\sigma = \rho W^2 r^2 \tag{15}$$

Thus, from Eq. (15),

$$\mathrm{E_{volumemax}} = \frac{1}{2}\sigma_{max} \tag{16}$$

where σ_{max} is the maximum tensile strength of the material (rim). For maximum kinetic energy per volume of the flywheel will depends on high tensile strength. Therefore, the choice of the maximum value of E_{volume} or E_{mass} will depend on the area of application. The maximum energy density ($E_{mass(max)}$) with respect to mass is given by Eq. (17).

$$\mathrm{E_{mass(max)}} = \frac{1}{2}\frac{\sigma_{max}}{\rho} \tag{17}$$

The general expression for the maximum energy density ($E_{mass(maxg)}$) of a flywheel with respect to mass is given in Eq. (18).

$$\mathrm{E_{mass(maxg)}} = \frac{K\sigma_{max}}{\rho} \tag{18}$$

where K is the shape factor. The values of K for several flywheel shapes are provided in **Table 5**.

4.2 Components of a flywheel energy storage

FES is made up of several parts namely motor-generator system, a motor control system, bearings, a flywheel, and a flywheel housing [35]. The type of bearings used in

Flywheel shape	Shape factor (K)
Constant stress disc	0.931
Flat unpierced	0.606
Thin rim	0.500
Rod or circular brush	0.333
Flat pierced disc	0.305

Table 5.
Flywheel shape factors [51, 52].

a flywheel is very important because of mechanical friction which is responsible for loss of energy in flywheel energy storage. Mechanical bearings are not ideal for FES because of the constant need for lubrication and maintenance. In modern FES, mechanical bearings are replaced with magnetic bearings. These can levitate the shaft thereby reducing the impacts of friction and the need to lubricate [35]. The motor-generator unit of a FES has a dual function. It can operate as a motor during the time of excess production of electricity in order to gain kinetic energy. When there is a need for electricity, it acts as a generator and converts the kinetic energy stored in the flywheel to electrical energy [51, 53]. FES housing provides a vacuum for the placement of FES. The housing guarantees that there is no loss of energy through air friction.

4.3 Advantages and drawbacks of FES

The merits of FES over other mechanical energy storage technologies include low costs of maintenance, very high efficiency, high power density, and long lifetime [38]. Its disadvantages are low energy densities and very high losses due to friction [54].

4.4 New advances in FES technology

There are several advances in FES technology geared towards reducing energy losses, improving efficiency, and widening the scope of applications.

Among these innovations is the use of high-temperature superconducting (HTS) bearings. This has the capacity of improving the overall round-trip efficiency of FES technology to over 90% according to [39].

In the area of applications, FES could be deployed in hybrid systems comprising of fuel cells, flow cells, ultra-capacitors, lithium-ion batteries, and small co-generation systems with low temperatures according to [43]. Besides, other new areas of applications of FES systems are in Kinetic Energy Recovery System [45], and in electric vehicle propulsion system [46], KERS/ERS.

5. Gravity energy storage (GES)

In GES electric pumps are used to pump water under a movable rock piston and through that, the rock mass is lifted. In the time of low production of electricity, the water which is already under high pressure from the rock mass is released to a turbine for the generation of electricity via a generator. Through this method, large quantities of water can be stored for several hours between 6 and 14 hours and can be made available to produce electricity when the needs arise. The size of the storage could be chosen between 1 to 10GWh and the rock piston diameter should be at least 100 meters.

6. Applications of mechanical energy storage systems in power system grid

With the increasing penetration of renewable energy sources in the grid network and the variability of these energy sources, it becomes necessary to bring a balance between power generation and demand. Therefore, energy storage systems integration into the grid becomes absolutely necessary [29]. The applications of mechanical energy storage systems in smart grid could be divided into energy-based and power-based applications.

6.1 Energy-based applications

Sufficient storage capacity is a requirement for energy-based applications to participate in very long discharges in a time window of one or more hours. PHS, CAES, and GES are used for energy-based applications discussed in the subsequent sub-sections.

6.1.1 Load following

Load following is a service in which the energy storage technology e.g., CAES, provides power for a long-time range (one or more hours) [16]. The intermittency of renewable energy sources accounts for the imbalance between generation and loads leading to variations in voltage and frequency. Mechanical energy storage systems such as PHS, CAES, and FES can provide the needed power to compensate for imbalance and stabilize the system frequency and voltage.

6.1.2 Peak-shaving

In this service, mechanical energy storage technologies, such as PHS, CAES, and GES are used to store energy during the time of excess production of power and to inject back energy into the grid during limited generation of power. In this service, power is delivered by the storage technology for several hours.

6.1.3 Transmission line curtailment

The capacity of a transmission line determines the optimal flow of power through it. When this exceeds the line capacity, the generation must be curtailed. In transmission line curtailment, CAES/PHS technology is used to inject power into power into the networks in a time window of 5–12 hours in line with the transmission line capacity. The applications of CAES for transmission curtailment are examined in [55–57].

6.1.4 Unit commitment

Owing to the uncertainties concerning variations in wind and of solar irradiation, managing the commitment of wind turbines and solar panels to meet the estimated demand always is difficult. Therefore, the use of energy-based storage system such as PHS in the networks may be useful to combat the effects of uncertainties in wind forecasting and to reduce the energy reserves if the system during its normal operation. In [58], the unit commitment problem was formulated in a power system with wind generation and CAES.

6.1.5 Spinning reserve

Mechanical energy storage systems such as PHS, CAES and GES can be used to compensate for unexpected contingencies for example the failure of a generating unit.

6.2 Power-based applications

In this application premium is placed on mechanical energy storage being able to charge or discharge within a very short interval of time (in milliseconds of time). FES is the best type of mechanical energy storage system for power-based applications because of its very short response time. Other energy storage systems that can be used for power-based applications include battery energy storage systems, [BESS], super-capacitors, and superconducting magnetic energy storage system (SMESS) [50]. The following subsections discusses some of the power-based applications where FES and other non-mechanical energy storage systems (such as BESS, super-capacitors and SMESS) could effectively be deployed.

6.2.1 Voltage control support

For the maintenance of proper voltage levels in the power system grid, the regulation of the flow of reactive power into power system network is very important. FES can effectively be used for both active and reactive power control thereby providing an excellent voltage control in the grid network.

6.2.2 Provision of inertia support

With the replacement of the fossil-fuel based power plants with the non-synchronized renewal energy power plants in the modern power grid, the total synchronized inertia of the system is getting diminished. This, if not checked, could result in frequency instability of the power system during contingency events. Mechanical energy storage systems especially FES (due to their short response time) can be used to emulate the provision of inertia of synchronous -based generators.

6.2.3 Fluctuation suppression

Certain loads in power systems (like electronic devices) are highly sensitive to non-sinusoidal voltage and current characteristics. FES may be used to inject power of high quality by quickly charging and discharging thereby smoothing out very short-term fluctuations.

6.2.4 Provision of a short duration power quality

Mechanical energy storage system especially FES can be deployed for the provision of short-duration power quality by supplying active power for very short duration in the range of 1–10 seconds.

7. Managing the high cost of mechanical energy storage systems

Energy storage systems especially PHS, CAES, and FES have been identified as a key device for realizing the goal of having high renewable penetration (wind and solar photovoltaic) in the modern grid. However, the extremely high cost of energy storage systems can constitute a barrier to achieving the above-mentioned goal. One way towards overcoming the challenge of high cost of energy storage systems is by the implementation of hybrid energy storage system. This involves the integration of different energy storage technologies for the implementation of several functions in the network. To achieve this, the technical characteristics of the different technologies such as the power and energy capacities, response time, and discharge time should be known. Besides, the application technical features should be analyzed for the proper hybridization of different technologies. **Table 6** shows the summary of application technical characteristics. Considering the various possible combine applications listed in section four of this work and referring to **Table 5** of applications technical characteristics, PHS and FES can be integrated together to provide for both power and energy-based applications. PHS can manage the energy-based applications such as peak shaving, and capacity firming while the FES can provide frequency excursion suppression, grid angular stability and grid voltage stability. Also, a black start application, frequency stability, regulation control and fluctuation suppression can be implemented by combining PHS and FES.

A second approach towards managing the extreme high cost of energy storage systems is by implementing the multi-functional utilization of the storage systems. In this approach, a single energy storage system is controlled in such a way to execute several functions. This will improve the cost effectiveness of energy storage system and will reduce the significant slack period of the storage system. However, the implementation of energy storage for multi-functional utilization will require the development of appropriate control methodologies. Without these, it will be impossible to utilize energy storage for multi-purpose applications.

Besides, the choice of suitable storage technology is very crucial for the multi-functional operation of an energy storage system. Some storage technologies are

Applications	Required time response	Reference duty cycle	ESS power (MW)	ESS AC voltage (kV)	Full power discharge duration
3-hour load shift	10 minutes	scheduled 3 hours of discharge	1–200	4.2–115	3 hours
10-hour load shift	10 minutes	Scheduled 10-hour discharge	1–200	4.2–115	10 hours
Renewable time shift	1 minute	Optimized by technology	2–200	4.2–34.5	5–12 hours
Fluctuation suppression	20 milliseconds	Continuous cycling	2–50	4.2–34.5	10 seconds
Short duration power quality	20 milliseconds	Hot standby	1–50	4.2–34.5	5 seconds
Long duration power quality	20 milliseconds	Hot standby	1–50	4.2–34.5	4 hours
Frequency excursion suppression	20 milliseconds	Hot standby	10–500	4.2–750	15 minutes
Grid frequency support	20 milliseconds	Hot standby	2–200	4.2–34.5	10–30 minutes
Angular stability	20 milliseconds	Hot standby	10–500	4.2–750	1 second
Voltage stability	20 milliseconds	Hot standby	10–500	4.2–750	1 seconds
Transmission curtailment	1 minute	Optimized by technology	2–200	4.5–34.5	5–12 hours

Table 6.
Application technical characteristics [23].

energy based, capable of delivering power over a prolonged period (e.g., PHS, CAES, etc) while others are power based (i.e., FES, SMESS), only being able of delivering high impulse power for few seconds. Therefore, it is necessary to evaluate the technical features of the various storage technologies to make appropriate choice of the required storage system.

Finally, good sizing methodology must be developed, bearing in mind that when energy storage systems are undersized, the reliability of the system becomes impaired while over sizing of the energy storage systems may results in less cost effectiveness.

8. Conclusion

In this work, a study on mechanical energy storage technologies and their various applications in the grid networks are presented. Their operating principles, topologies, various subsystems, performance measures, advantages and drawbacks were discussed.

In addition, the work also presents the development of output power equations for each mechanical energy storage type based on the fundamental principles of potential energy due to gravity, air compression atvery high pressure, and kinetic energy of rotating masses.

Moreover, the capacity of energy storage systems in providing flexibility that is essential for robust, high quality stable electrical systems, in a grid network with a high penetrations of renewable energy sources were examined in detail.

Further highlights in the work include key approaches to improving the efficiency of mechanical energy storage systems especially compressed air energy storage system and flywheel energy storage system. Such strategies take in cognisance the need of improving the performance of every subsystem and the deployment of new advances in technology such as the use of high temperature superconducting bearings for FES.

Lastly, the management of the high cost of energy storage systems through the implementation of hybrid energy storage systems, multi-functional applications of energy storage technologies and appropriate sizing methodologies were discussed. The diligent pursuit of these strategies will surely reduce the costs factor in energy storage systems and consequently result in more of their deployment into the grid networks.

Author details

Chukwuemeka Emmanuel Okafor* and Komla Agbenyo Folly
University of Cape Town, Cape Town, South Africa

*Address all correspondence to: okfchu002@myuct.ac.za

References

[1] Nadeem F, Suhail S, Pranshant KT. Comparative review of energy storage systems, thier roles and impacts on future power systems. IEEE Access. 2019;7:4555-4585

[2] Prantil V, Decker T. The Captains of Energy. Manhattan, New York City: Springer; 2015

[3] Boicea V. Energy storage technologies: The past and the present. Proceedings of IEEE. 2014;**102**(11):1774-1794

[4] Rehman S, Luai M, Hadhrami A, Mahbub M. Pumped hydro energy system: A technological review. Renewable and Sustainable Energy Reviews. 2015;**44**:586-598

[5] Dunn B, Kamath K, Tarascon J. Electrical energy storage for the grid. Materials for Grid Energy. 2011;**334**: 928-934

[6] Chen DHS. Progress in electrical energy storage system: A critical Review. Progress in Natural Science. 2009;**19**: 291-312

[7] R. Carnegie, D. Gotham, D. Nderitu and Preckel P. Utility Scale Energy Storage Systems: State Utility Forecasting Group, West Lafayette, Indiana: 2013

[8] Mohamad F, Teh J, Ming Lai C, Chen L. Development of energy storage systems for power network reliability: A review. Energies. 2018;**2278**:1-19

[9] Johnson S, Papageorgiou D, Harper M, Rhodes J. The economic and reliability impacts of grid-scale storage in a high penetration renewable energy system. Advances in Applied Energy. 2021;**23**:1-19

[10] Palizban O, Kauhaniemi K. Energy storage in modern grids-Matrix of technologies and applications. Journal of Energy Storage. 2016;**6**:248-259

[11] Miller M, Viernstein L, Nam Truong C. Evaluation of grid-level adaptability for stationary battery energy storage system applications in Europe. Journal of Energy Storage. 2017;**9**:1-11

[12] Barton J, Infied D. Energy storage and its use with intermittent renewable energy. IEEE Transactions on Energy Conversion. 2004;**19**:441-448

[13] Mears LD, Gotschall HL. Epri-Doe Handbook of Energy Storage for Transmssion and Distribution Applications. Washington, DC: U.S Department of Energy; 2003

[14] Dusonchet S, Favuzza S, Massaro F, Telaretti E, Zizzo G. Technological and legislative status point of stationary energy storages in the EU. Renewable and Sustainable Energy Reviews. 2019; **101**:158-167

[15] Yao L, Yang B, Zhuang J, Xue J. Challenges and progresses of energy storage technology and its application in power systems. Journal of Modern Power Systems and Clean Energy. 2016;**4**:519-528

[16] Shaqsi A, Sopian K, Hinai A. Review of energy storage services, applications, limitations and benefits. Energy Reports. 2020;**6**:288-308

[17] Zidar M, Georgilakis PS, Hatziargyriou ND, Capuder T. Review of energy storage allocation in power distribution networks: Applications, methods and future research. IET Generation Transmission and Distribution. 2016;**10**:645-652

[18] Diaz-Gonzalez A, Gomis-Bellmunt O. Energy Storage in Power Systems.

West Sussex: John Wiley & Sons Ltd.; 2016

[19] Molina MG. Energy storage and power electronics technologies: A strong combination to empower the transformation to the smart grid. Proceedings of IEEE. 2017;**105**:2191-2219

[20] Stecca M, Elizondo LR, Soeiro TB, Bauer P. A comprehensive review of the integration of battery energy storage systems into distribution networks. Industrial Electronics Society. 2020;**1**:46-65

[21] Hess HC, Schimpe M, Kucevic D, Jossan A. Lithium -Ion battery storage system design tailored for applications in modern powe grids. Energies. 2017;**10**:1-42

[22] Wei W, Wu D, Wang Z, Mei S, Catalao J. Impact of energy storage on economic dispatch of distribution systems: A multi-parametric linear programming approach and its implications. Power and Energy. 2020;7: 243-253

[23] Jung J.. 21 December 2023. [Online]. Available: http://www.jume.vfrgs.bu/b itstream/handle [Accessed: December 21, 2023]

[24] Arup J. Five Minutes Guide to Electricity Storage Technologies. London, UK: Arup; 2014

[25] Fortune Business insights, Energy and Power. 2022. [Online]. Available: h ttp://www.fortunebusinessinsights.com/ industry-reports [Accessed: December 21, 2022]

[26] Skoglund P. Large Scale Energy Storage. Umea: Umea University; 2017

[27] Paliban O, Kauhaniemi K. Energy storage systems in modern grid- matrix of technologies and applications. Journal of Energy Storage. 2016;**6**:248-259

[28] Carija Z, Kranjeevic I, Banic V, Cavrak M. Numerical analyses of wells turbine for wave power conversion. Engineering Review. 2012;**32**(3):141-146

[29] DOE. Global Energy Storage Database. Washington, DC: US Department of Energy; 2017

[30] Wang C, Wu Z, Yang X. Modeling and verification of hybrid energy storage system basedo on micro compressed air energy storage. Autum Electric Power system. 2014;**38**:22-26

[31] Wang C, Chen L, Liu F. Thermal - Wind storage joint operation of power system considering pumped storage and distributed compressed air energy storage. In: Power System Computation Conference. Poland; 2014

[32] Why energy storage technologies (Pumped Hydropower) [online]. Available from: https://energystorage.org/why-energy-storage/technologies/pumped-hydropower. Accessed: January12, 2021]

[33] Paul B. Power System Energy Storage Technologies. Amsterdem, The Netherlands: Elsevier; 2018

[34] Madlener R, Latz J. Centralized and Decentralized compressed air energy storage for enhanced grd integration of wind power. In: Institute for Future Energy Consumers Needs and Behavior. Aachan, Germany; 2009

[35] Alzola P, Sabastian R, Quesada J, Colmener A. Review of flywheel based energy storage system. In: International Conference on Power Engineering, Energy and Electric Drives. Malanga; 2011

[36] Bolund B, Bernhoff H, Leijon M. Flywheel energy and power storage system. Renewable and Sustainable energy. 2014;**11**:235-258

[37] Chen L, Zheng T, Mei S, Xue X. Review and prospects of Compressed air energy storage system. Power System clean energy. 2016;**4**:529-541

[38] Berhoff H, Leijon M, Bolund B. Flywhell energy and power storage systems. Renewable and Sustainable energy Journal. 2007;**11**:235-258

[39] Zhang C, Tseng K. Design and control of a novel flywheel energy storage system assisted by hybrid mechanical-magnetic bearings. Mechatronics. 2013;**23**(3):297-309

[40] Bankston S, Changki M. Geometry modification of flywheel and its effects on energy storage. Energy Research Journal. 2015;**6**:54-63

[41] Dain T. Evaluation of energy storage technologies for integration with renewable electricity,quantifying experts opinion. Science Direct. 2012;**3**: 29-49

[42] Ter-Gezarian. Energy Storage for Power System. London: Peter Peregrinus; 2011

[43] Danbone Sessa S, Tortella A, Andriollo M, Benato R. Li-ion Battery-Flywheel hybrid storage system: countering battery aging during a frequency regulation service. Applied Science. 2018;**2018**:1-15

[44] Daneshi H, Daneshi A, Tabari N, Jahromi A. Security-constrained unit commitment in a system with wind generation and compressed air storage. In: 6th International Conference on the European Energy Market. Belgium; 2009

[45] Meishner F, Sauer D. Wayside energy recovery systems in DC urban railway grids. eTransportation. 2019; **2019**:1-20

[46] Sliwinski C. Kinetic energy recovery systems in motor vehicles. Materials Science and Engineering. 2016;**2016**:1-11

[47] Guney M, Tepe Y. Classification and assessment of Energy Storage Systems. Renewable and sustainable energy journal. 2016;**13**:1934-1945

[48] Deoloitte Centre for Energy Solutions. Electricity storage technologies, impacts,and prospects. 2015

[49] Crotogino F, Mohmeyer K, Scharf R, Bremen E. Huntorf CAES: More than 20 years of Successful Operation. California, USA: Solution Mining Research Institute; 2001

[50] La Monica M. Compressed air energy storage makes a come back. In: Spectrum IEEE. New York City, New York. 2013

[51] Wilson A. Machines, power and the ancient economy. The Journal of the Roman studies. 2012;**92**:1-32

[52] XIiaojun I, Bahareh A, Alan P, Zhiyang W, Hamid T. A utility scale flywheel energy storage system with a shaftless, hubless high resistance steel. IEEE Transactions on Industrial Electronics. 2012;**68**:6667-6678

[53] Aneke M, Wang M. Energy storage technologies and real-life applications: A state-of the art review. Applied Energy. 2016;**179**:350-377

[54] Pena Alzola R, Sebastian R, Quesada J, Colemenar A. Review of fly wheel based energy storage system. In: International Conference on Power

Engineering, Energy and Electric Drives. Malaga, Spain; 2011

[55] Anagnostopoulos J, Papantomis D. Simulation and size optimization of a pumped-storage power plant recovery of windfarm renewable energy. Renewable Energy. 2008;**33**:1685-1694

[56] Denholm P, Sioshansi R. The value of compressed air energy storage with wind in a transmission constrained electric power system. Energy Policy. 2009;**37**:3149-3158

[57] Dursun S, Alboyaci B. The contribution of wind -hydro pumped storage systems in meeting Turkey's electric energy demand. Renewable and Sustainable Energy Reviews. 2010;**14**: 1979-1988

[58] Wicki S, Hansen E. Clean energy storage technology in the making: An innovative systems perspective on flywheel energy storage. Journal of Clean Energy Production. 2017;**162**:1118-1134

Chapter 2

A Review on Superconducting Magnetic Energy Storage System Applications

Narges S. Ghiasi and Seyyed Mohammad Sadegh Ghiasi

Abstract

Superconducting Magnetic Energy Storage is one of the most substantial storage devices. Due to its technological advancements in recent years, it has been considered reliable energy storage in many applications. This storage device has been separated into two organizations, toroid and solenoid, selected for the intended application constraints. It has also been used in many industries, such as transportation, renewable energy utilization, power system stabilization, and quality improvement. This chapter discusses various SMES structures and their applications in electric and power systems. Here, the authors try to deliver a comprehensive view for scholars whose research is related to the SMES by examination of the published articles while providing a brief guideline of this modern technology and its applications.

Keywords: energy storage, SMES, electrical power systems, energy systems, renewable energy resources

1. Introduction

Today, many Energy Storage Systems (ESS) are being used. Users have various options according to the application and parameters such as cost, available room, accuracy, lifetime, and efficiency. Among numerous ESS technologies, Battery Energy Storage Systems (BESS), Super Capacitor Energy Storage Systems (SCES), Flywheel Energy Storage Systems (FESS), Compressed Air Energy Storage Systems (CAES), and Superconducting Magnetic Energy Storage Systems (SMES) are the leading viable technologies. Each of these technologies has strengths and weaknesses. The negative attributes of BESS are limited lifecycle, failure of deep discharge, and processing of lead afterward. Using toxic heavy metals is another issue with some types of BESS [1]. The drawbacks of SCES are a limited range of operating voltage, limited energy output in fast cyclic operation [2], and toxic and corrosive materials [3]. As the limitations of FESS, the possibility of mechanical failure and dissociation [3], considerable standby losses [4, 5], the dependence of stored energy on magnetic sources (usually permanent magnet), and the deterioration of magnetic sources and consequently in energy storage capacity [6] can be mentioned. Furthermore, in some specific applications, such as vehicular applications, increasing the center of gravity height of the vehicle followed by unbalancing issues on the vehicle is problematic [7]. Slow response [8], low round-trip efficiency [1, 9, 10], limitations imposed by topographical conditions, and negative environmental impact [9–11] are the main disadvantages of the CAES system. Quick positioning time (reaction time plus rising to peak discharge power), rapid charging time, considerable capacity, high cycle efficiency, instantaneous power output, reliability, no self-discharge, and low maintenance are some listed benefits of SMES [11–14]. Two major concerns raised to SMES technologies are high cost and strong magnetic force due to the high magnetic field [15]. As mentioned in [16], by improvement in the superconductor manufacturing industry and the downward trend of high-temperature conductors' cost, SMES technology will become an economical and available storage device [17–20]. Besides, there are many ongoing types of research to lessen cooling system costs. Moreover, in some cases, using SMES helps decrease long-term costs. To answer the environmental impact of the magnetic field, the authors propose a Force-Balanced Coil (FBC) in [21, 22].

In this chapter, while briefly reviewing the technologies of control systems and system types in Section 2, Section 3 examines the superconducting magnetic energy storage system applications in the articles related to this technology. Also, the conclusion section is advanced in the fourth section.

2. System type

The SMES system is a DC device that keeps the energy in a magnetic field. The current flows through an inductor kept in specific conditions providing superconductivity; thus, a strong magnetic field is created. The superconducting coil unit, power conditioning, and cryogenic subsystem are the three main parts of a typical SMES. The Power Conditioning System (PCS) plays a crucial role in power exchange between the superconducting coil and the AC system. According to the configuration of this subsystem, there are three types of SMES: thyristor-based SMES, Voltage Source Inverter-based SMES (VSI-SMES), and Current Source Inverter-based SMES (CSI-SMES). Regardless of the kind of SMES control system, its applications are mentioned in the following section.

From the structural viewpoint, there are two types of SMES: Toroidal and Solenoidal. Considering the application, investment, available room, production availability, etc., the structure of the SMES system in the designing step should be selected. The solenoid type is more convenient to build and can also manage mechanical stresses due to less wire consumption. Compared to the toroidal type, it is more cost-effective. Despite the solenoidal geometry, the toroidal SMES has less stray field and decreases the vertical component of the magnetic field on the conductor; therefore, it can be expected that the necessities of the materials and AC losses are overshadowed [17, 23, 24]. **Figure 1** shows both solenoidal and toroidal structures of the SMES.

3. SMES application

The specific characteristics of a superconducting magnetic energy storage system provide outstanding capabilities making it a fitting choice for many applications. Applications of SMES are defined in the following subsections by mentioning many cases in which its effectiveness in power systems has been proven. This section has made an effort to provide a directory of SMES technology.

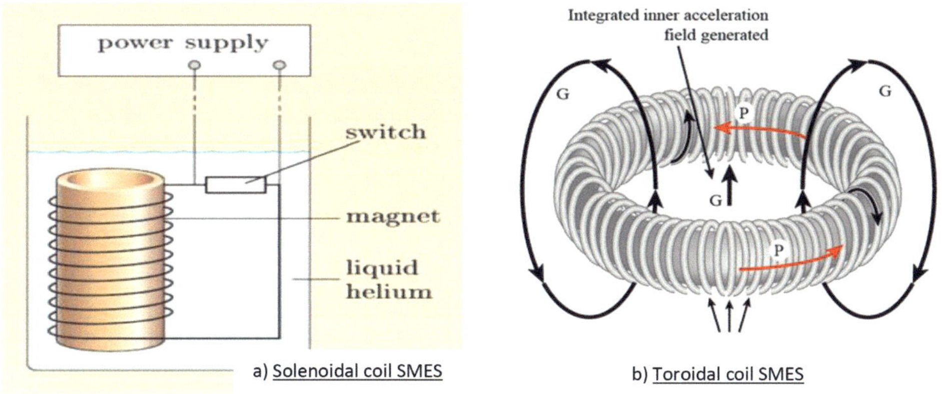

Figure 1.
Different types of SMES structure: (a) solenoidal geography, (b) toroidal geography.

3.1 Power quality improvement

The end-user sensitive loads require undistorted power. Supplying the users at an acceptable voltage range and mitigating disturbances are critical tasks. SMES features such as fast response, large capacity, and ability to control active and reactive power simultaneously make it a suitable option for a power quality boost [8–10]. Mitigating the voltage sag using the stored energy in SMES helps recover the standard voltage range in less than a second, leading to power quality improvement, which is essential for sensitive loads [25–36].

3.2 Power fluctuations compensation

Besides all their benefits, renewable resources may cause unstable power due to the uncertainty of renewable resources. Wind speed variations in a wind farm and sun radiation variations in the case of using PV cells as the power generating system result in problems such as lump flicker, timing device imprecision, and shortened hardware life cycle. Considering its fast response, the SMES unit can compensate for the differences between demand and generation by absorbing or releasing power, minimizing the adverse effects on the demand side. The fast response characteristic of SMES has been proven effective in reacting to the power difference between both sides, smoothing the power fluctuations [37–39]. In Ref. [40–44], different controlling methods are employed due to various resources and load sensitivity.

3.3 Power oscillation compensation

Load changes and system faults cause low-frequency power oscillation between 0.5 and 1 Hz. Considering the probability of such disturbances, it seems the system's oscillation damping is essential that should be applied rapidly. As a storage device capable of exchanging a large amount of active and reactive power simultaneously with the system in the minimum time, SMES is introduced as an efficient power compensator [32]. With the proper controller design, SMES could compensate for system's oscillation under light and heavy load flow and renewable resources uncertainties [45–49].

3.4 Stability

Power system's stability includes voltage, power angle, and frequency stability. To provide these parameters stability, generator excitation control is established to be an effective solution. Still, more is needed due to power networks' rapidly increasing size and complexity. SMES is one of the most encouraging members of the FACTS collection, which can supply/receive active and reactive power immediately and has been highly projected as one of the most effective controllers of power system stabilization. The concept of its operation is simple: a superconducting coil that stores or releases energy by charging or discharging a thyristor-controlled power converter connected to an AC power system. The firing angles of thyristor variation in an appropriate mode determine the course of energy transmission. This process helps to mitigate the released energy during a disturbance and brings back stability to the system [50–71].

3.4.1 Dynamic stability

One of the problems of the isolated power system is the fluctuation and instability caused by abrupt alterations in load or generation unit followed by the grid's deviation of frequency and power. Power system oscillations happen when conflicts such as sudden burden changes occur. The system's suppressing must be so that the synchronous generators can return to steady-state conditions. Particularly when the end of the transmission line is experiencing sudden load distress, the generators must have incessant control to suppress oscillations in the system. Many proceedings have been proposed to fortify the damping process, including power system stabilizers, optimal control of the turbine-governor system, and static phase shifter. Successful tests of the BPA 30 MJ unit and superconductive magnetic energy storage (SMES) systems have gained scholars' attention in power applications. Although the device's original resolution in that experiment was load leveling, another function of the SMES unit was to enhance the system performance by providing precise power modulation in a dynamic period [72–77].

3.5 Shaft oscillation minimization

Despite all assumptions analyzing the power system dynamic performance based on the system integrity, the turbine-generator rotor has built-in a complicated mechanical structure and consists of several predominant masses. The generator under tension leads to twisting stress among the shaft components in a sub-synchronous range which may cause electrical system damage. On the other hand, some disturbances in the electrical system might have the same effect on shafts, so the life expectancy of the generator would decrease. Given the mentioned complications, a proper damper should be applied. Using SMES associated with a controlling system will be helpful as a compensating system as the SMES can compensate for the system fault leading to less electrical tension and, therefore, less mechanical stress on the generators [78, 79].

3.6 Voltage stability

Wind generators, especially squirrel cage induction generators, consume a noticeable amount of reactive power. Moreover, voltage sag induced by a sudden increase in loads is a severe problem in the electrical network. As a result, reactive power control is consequential in maintaining the standard voltage level in the system. A power compensation device should be employed to compensate for the voltage sag. SMES is a capable source due to the rapid charge and discharge periods. Connected to a power system with a power electronic converter, SMES can convey energy into the power system in milliseconds, resulting in maintaining load voltage in a normal range, improving the network stability, and compensating voltage sag properly [80–92]. It should be noted that locating SMES units in the system should be carried out based on the quantitative voltage stability index [93].

3.7 Sub-synchronous resonance compensation

Sub-synchronous resonance usually occurs in power systems containing steam generators and compensated lines by parallel capacitors, which cause shaft resonances and consequential damages. When a disturbance happens in these systems, in the case of the sub-synchronous resonance effect, the shaft rotates at sub-synchronous frequencies along with the base synchronous speed, which, if not confined, results in a shaft break. The sub-synchronous resonances are the effects of improper energy exchanges between turbo-generator and power systems, and they could originate anywhere on the system. The capability of rapidly absorbing and delivering energy has made SMES an appropriate choice for compensating for these disturbances to prevent sub-synchronous resonances [74, 94–98].

3.8 Renewable energy resources

Renewable energy generation is widely used to address energy shortages and environmental problems. Since the output power of the solar cells and the wind turbines changes with the variation of sunlight irradiation and the wind speed and high level of the renewable resources penetration into the power systems can trigger instability, unreliability, and power quality issues, the system must be able to withstand current and reactive power fluctuations, otherwise, the distortion causes harms to the system and critical loads. To cope with these problems, using ESS in renewable power combinations has been greatly noticed. In this case, the performance criteria are energy storage capacity, power output, and life cycle. To alleviate the mentioned issues, SMES can be applied, which can charge and discharge immediately. In other words, it can absorb high quantities of power in the time of lack of demand and deliver high quantities of power in the time of lack of production of electric power. Therefore, the utilization of renewable energy resources, despite their intermittent nature, is associated with more accessibility, and the compensation is applied in less than a minute or some cases, a few minutes [82–85, 99–152].

3.9 Frequency control

The system frequency is a dependent factor on the active power balance of that system. Since frequency is a common parameter in the system, the alternation of the active power in one point reflects on the whole system. Governor is not capable of absorbing the frequency fluctuation in a short time. Because of that, using SMES as storage with quick response time while supplying high power density has been a practical solution in many systems to compensate for the sudden load change and, consequently, frequency control [111, 153–166].

3.10 Automatic generation control (AGC)

Failure to adapt demand and generation in the power system causes unsteady operation and disrupts the system's dynamic performance. Adding a storage device to the system will cause the AGC to achieve its goal of eliminating this disorder faster, improving the transient state operation. Due to features such as a high-density discharge rate, the minimum time required for power flow reversal, and low maintenance requirements, many systems switch to SMES, which is developed to control active and reactive power simultaneously. When an increase step change in power demand occurs, the energy stored in the SMES is injected through the PCS as an alternative to the power system. When the governor and controllers start and retrieve the position, SMES returns to the initial value of the current. When an unexpected drop in power demand occurs, the SMES intervenes and absorbs part of the excess energy. After recovering the position, this absorbed energy is released, and the SMES returns to its standard value [167–170].

3.11 Low voltage ride-through (LVRT)

A neighboring grid fault causes a drop in the grid voltage at the junction point of the generator and network; low voltage ride-through happens. This situation limits the power flow. If there is a considerable input power, the power inequity leads to a surge in the back-to-back converter's turbine speed or DC bus voltage. An ESS could avoid over-speeding of the turbine and even out the DC bus voltage level. High power ability and instant response are essential for this application; thus, SMES is a great storage device to fit [171, 172]. The SMES can restrict the fault current while compensating for voltage drop and eliminate the residual power fluctuation caused by the disturbance [173, 174].

3.12 Protection

Installation of SMES alongside power end-users as a SMES-based UPS can protect critical loads during faults. SMES alone does not affect fault current limitation at ground fault location but, combined with SFCL, plays an influential role in an error. This ESS can retain the network voltage at or near its nominal value in case of a fault. Generator tripping is one of the most critical emergency controls to prevent damage to the power system. Considering the availability of SMES to provide long-time energy or high-density power for a load, it could protect a specific load alongside a proper controller. Therefore, in a system with prioritized loads, the ESS could protect the sensitive loads at any given time. The current flowing through the coil is defined based on the operating point of the given load and the time of protection [175–178].

3.13 Black start

A power loss over a large geographical area for a noticeable period, called a blackout, is a severe threat to the network. Besides all the effects on the residential consumers, the industries face irrecoverable losses at this time. Therefore, power system restoration or black start is vital for a SMES unit to cope. Unlike traditional methods, utilizing SMES has advantages such as faster start speed than other generation methods, multipath black start, and more environmentally friendly than other black start units. Considering the four-quadrant operation, SMES can deal with the primary problems of the black start process, which are over-voltage, voltage drop, and power oscillation [66].

3.14 Load leveling

Load leveling is a concept that helps with having a balanced power profile by decreasing demand at peak times and filling the valleys (**Figure 2**). Due to the principles of load leveling, SMES has two advantages: High storage efficiency of 90% and no site limitation. In low demand time, SMES is recharged, and thus the generator operates in its optimal range (filling the valley), and at the peak-demand time, the SMES is discharged. The difference between the demanded and generated power (in the optimal range) is compensated by discharging its power in the network [171, 179, 180].

On the other hand, every generator has an optimum work point, achieving that, parameters such as generator lifetime, total loss, and system overall performance would be at the prime mode. On the other hand, consumption would not be monotonic and uniform, and the profile consists of peaks and valleys. Concerning the discrepancy between generation and consumption curves, a SMES unit could be utilized to preserve the optimum system performance and prevent waste of energy. Receiving energy during the underload situation and releasing it during the overload situation, the SMES charged and discharged respectively; hence the power would have a smooth curve. Having a large capacity and fast response time, SMES is capable of rapid load following to develop the invariant outage for the generating unit [119].

3.15 Spinning reserve

Spinning reserve is defined as the dormant capacity that the system operator could employ. Considering the capacity and response time, a SMES unit represents a significant spinning reserve capacity that could quickly be activated by Primary Frequency Control (PFC) and improve system parameters [10, 181].

3.16 Transportation

The transportation application can be investigated from three points of view. First, considering the required power while charging, electric vehicles draw significant

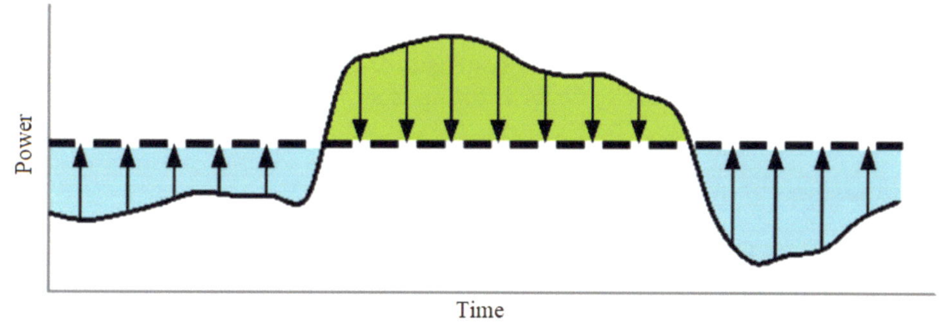

Figure 2.
Load leveling in the power profile.

current from the power system, which causes instability and following complications. The primary role of stationary SMES in road vehicles and railway transportation systems is to supply a high-frequency component [105, 106]. Second, as an ESS, the SMES is an auxiliary unit causing more contribution of Renewable Energy Resources (RES) as the energy supplier for electrical transportation, whether rail or road. Although renewable power stations such as wind farms or PV cells can provide the necessary power to run mentioned transportation systems, due to variations of output power subordinate to weather conditions, an energy storage system is required to supply the load with standard quality. During low-demand periods and peak times, if power generation cannot support all the consumers, a previously stored amount of power is released into the grid, which is feasible by utilizing an ESS with a fast response [121, 123]. Third, in railway systems, a tremendous amount of energy is wasted during deceleration in the form of thermal energy. Retrieving and returning this energy to be used in the system saves energy and reduces costs in the long term. Large capacity and fast response make SMES an excellent candidate for absorbing energy during acceleration and releasing it during deceleration [182–187].

3.17 P & Q control

SMES can absorb or deliver active and reactive powers independently and simultaneously. Although the output of active power from SMES relies on the energy stored in the coil, SMES can uninterruptedly function throughout its reactive power range to regulate the voltage of the common coupling point. The converter firing angle controls the domains of active and reactive power [188–192].

3.18 Reliability

To strengthen the reliability of the power systems, storage systems should cope with electrical outages caused by natural events and support demands when power system failures arise accordingly. The power and energy capacity of SMES provide reliability for the system by ensuring the power supply for a given load while observing the standard range for different parameters [55, 107, 108, 120].

3.19 Uninterruptable power supplies

Widespread usage of electric devices from domestic consumers to industry consumers emphasizes the continuous quality of supply. Some equipment demands an uninterrupted electricity supply among all facilities due to high prices or high-performance sensitivity. With high energy density and fast response, SMES can be a promising candidate to supply these loads for up to several hours [3, 10, 193].

4. Conclusion

In this chapter, based on previously published articles on the technology of SMES, the potential usages of this technology have been reviewed. Considering the application this storage system is used for, the type and the control strategy should be selected. This storage system has been proven effective for many industrial applications, such as active and reactive power control, system stability, and power quality.

Given its technology advancement, price reduction, and the required room, it is anticipated that the SMES has a clear horizon to be more involved in areas like reliability and the transportation industry. Due to the exhaustiveness of the study, this article can be used as an acceptable guideline for researchers, engineers, readers, and academicians working in the fields related to this technology.

Author details

Narges S. Ghiasi[1] and Seyyed Mohammad Sadegh Ghiasi[2*]

1 Holcombe Department of Electrical and Computer Engineering Clemson University, Charleston, United States

2 Department of Electrical Engineering, Ahvaz Branch, Islamic Azad University, Ahvaz, Iran

*Address all correspondence to: smsghiasi@aut.ac.ir

References

[1] Aneke M, Wang M. Energy storage technologies and real-life applications – A state of the art review. Applied Energy. 2016;**179**:350-377

[2] Morandi A, Trevisani L, Negrini F, Ribani PL, Fabbri M. Feasibility of superconducting magnetic energy storage on board of ground vehicles with present state-of-the-art superconductors. IEEE Transactions on Applied Superconductivity. April 2012;**22**(2):5700106-5700106. Art no. 5700106, DOI: 10.1109/TASC.2011.2177266

[3] Ries G, Neumueller HW. Comparison of energy storage in flywheels and SMES. Physica C: Superconductivity and its Applications. 2001;**357-360**(Suppl. 1): 1306-1310

[4] Kousksou T, Bruel P, Jamil A, El Rhafiki T, Zeraouli Y. Energy storage: Applications and challenges. Solar Energy Materials & Solar Cells. 2014;**120**(PART A):59-80

[5] Kyriakopoulos GL, Arabatzis G. Electrical energy storage systems in electricity generation: Energy policies, innovative technologies, and regulatory regimes. Renewable and Sustainable Energy Reviews. 2016;**56**:1044-1067

[6] Hasan NS, Hassan MY, Majid MS, Rahman HA. Review of storage schemes for wind energy systems. Renewable and Sustainable Energy Reviews. 2013;**21**:237-247

[7] Mahlia TMI, Saktisahdan TJ, Jannifar A, Hasan MH, Matseelar HSC. A review of available methods and development on energy storage; technology update. Renewable and Sustainable Energy Reviews. 2014;**33**:532-545

[8] Ferreira HL, Garde R, Fulli G, Kling W, Lopes JP. Characterisation of electrical energy storage technologies. Energy. 2013;**53**:288-298

[9] Chatzivasileiadi A, Ampatzi E, Knight I. Characteristics of electrical energy storage technologies and their applications in buildings. Renewable and Sustainable Energy Reviews. 2013;**25**:814-830

[10] Luo X, Wang J, Dooner M, Clarke J. Overview of current development in electrical energy storage technologies and the application potential in power system operation. Applied Energy. 2015;**137**:511-536

[11] Zhao H, Wu Q, Hu S, Xu H, Rasmussen CN. Review of energy storage system for wind power integration support. Applied Energy. 2015;**137**:545-553

[12] Chauhan A, Saini RP. A review on integrated renewable energy System based power generation for stand-alone applications: Configurations, storage options, sizing methodologies and control. Renewable and Sustainable Energy Reviews. 2014;**38**:99-120

[13] Tan X, Li Q, Wang H. Advances and trends of energy storage technology in microgrid. International Journal of Electrical Power & Energy Systems. 2013;**44**(1):179-191

[14] Rodrigues EMG, Godina R, Santos SF, Bizuayehu AW, Contreras J, Catalão JPS. Energy storage systems supporting increased penetration of renewables in islanded systems. Energy. 2014;**75**:265-280

[15] Akorede MF, Hizam H, Pouresmaeil E. Distributed energy

resources and benefits to the environment. Renewable and Sustainable Energy Reviews. 2010;**14**(2):724-734

[16] Jin JX, Chen XY. Study on the SMES application solutions for smart grid. Physics Procedia. 2012;**36**:902-907

[17] Yagai T, Mizuno S, Okubo T, Mizuochi S, Kamibayashi M, Jinbo M. et al. Development of Design for Large Scale Conductors and Coils Using MgB2 for Superconducting Magnetic Energy Storage Device, Cryogenics (Guildf). 2018;**96**:75-82. ISSN 0011-2275

[18] Mukherjee P, Rao VV. Physica C: Superconductivity and its applications design and development of high temperature superconducting magnetic energy storage for power applications - a review. Physica C: Superconductivity and its Applications. 2019;**563**(March):67-73

[19] Saranya S, Saravanan B. Effect of emission in SMES based unit commitment using modified Henry gas solubility optimization. Journal of Energy Storage. 2020;**29**:101380

[20] Zimmermann AW, Sharkh SM. Design of a 1 MJ/100 kW high temperature superconducting magnet for energy storage. Energy Reports. 2020;**6**:180-188

[21] Nomura S, Watanabe N, Suzuki C, Ajikawa H, Uyama M, Kajita S, et al. Advanced configuration of superconducting magnetic energy storage. Energy. 2005;**30**(11-12 SPEC. ISS):2115-2127

[22] Shi J, Liao M, Zhou X, Li Y, Zhang L, Liao M, et al. Integrate Method to alleviate the High Frequency PWM Pulse Voltage on SMES Magnet. Energy Procedia; 2019;**158**:4816-4821. ISSN 1876-6102. DOI: 10.1016/j.egypro.2019.01.714

[23] Morandi A, Fabbri M, Gholizad B, Grilli F, Sirois F, Zermeño VMR. et al. Design and comparison of a 1-MW/5-s HTS SMES with toroidal and solenoidal geometry. In: IEEE Transactions on Applied Superconductivity. June 2016;**26**(4):1-6. Art no. 5700606. DOI: 10.1109/TASC.2016.2535271

[24] Yi KP et al. A design methodology for toroid-type SMES using analytical and finite-element method. In: IEEE Transactions on Applied Superconductivity. June 2013;**23**(3):4900404-4900404. Art no. 4900404. DOI: 10.1109/TASC.2012.2233254

[25] Kim AR, Kim JG, Kim S, Park M, Yu IK, Seong KC, et al. A feasibility study on HTS SMES applications for power quality enhancement through both software simulations and hardware-based experiments. Physica C: Superconductivity and its Applications. 2011;**471**(21-22):1404-1408

[26] Rabiee A, Khorramdel H, Aghaei J. A review of energy storage systems in microgrids with wind turbines. Renewable and Sustainable Energy Reviews. 2013;**18**:316-326

[27] Díaz-González F, Sumper A, Gomis-Bellmunt O, Villafáfila-Robles R. A review of energy storage technologies for wind power applications. Renewable and Sustainable Energy Reviews. 2012;**16**(4):2154-2171

[28] Planas E, Andreu J, Gárate JI, Martínez De Alegría I, Ibarra E. AC and DC technology in microgrids: A review. Renewable and Sustainable Energy Reviews. 2015;**43**:726-749

[29] Chen XY, Jin JX. Application prospects of integrated SMES technology for future smart grids. In: 2013 IEEE International Conference on Applied Superconductivity and Electromagnetic

Devices. Beijing, China; 2013. pp. 517-518. DOI: 10.1109/ASEMD.2013.6780834

[30] Nitta T. Applied superconductivity for advanced electric power system. Physica C: Superconductivity and its Applications. 2001;**357-360**((Suppl. 1): 1245-1254

[31] Nguyen TT, Yoo HJ, Kim HM. Applying model predictive control to SMES system in microgrids for eddy current losses reduction. In: IEEE Transactions on Applied Superconductivity. 2016;**26**(4)1-5, June 2016, Art no. 5400405. DOI: 10.1109/TASC.2016.2524511

[32] Ren L, et al. Development of a movable HTS SMES system. In: IEEE Transactions on Applied Superconductivity. Aug 2015;**8223**(4):1-9. Art no. 5701109. DOI: 10.1109/TASC.2015.2437335

[33] Panda AK, Penthia T. Electrical power and energy systems design and modeling of SMES based SAPF for pulsed power load demands. International Journal of Electrical Power & Energy Systems. 2017;**92**:114-124

[34] Hall PJ, Bain EJ. Energy-storage technologies and electricity generation. Energy Policy. 2008;**36**(12):4352-4355

[35] Wang Z, Chau KT, Yuwen B, Zhang Z, Li F. Power compensation and power quality improvement based on multiple-channel current source converter fed HT SMES. In: IEEE Transactions on Applied Superconductivity. June 2012;**22**(3):5701204-5701204. Art no. 5701204. DOI: 10.1109/TASC.2011.2174573

[36] Seo HR, Kim AR, Park M, Yu IK. Power quality enhancement of renewable energy source power network using SMES system. Physica C: Superconductivity and its Applications. 2011;**471**(21-22):1409-1412

[37] Lin X, Lei Y, Zhu Y. A novel superconducting magnetic energy storage system design based on a three-level T-type converter and its energy-shaping control strategy. Electric Power Systems Research. 2018;**162**(24):64-73

[38] Saejia M, Ngamroo I. Alleviation of power fluctuation in interconnected power systems with wind farm by SMES with optimal coil size. In: IEEE Transactions on Applied Superconductivity. June 2012;**22**(3):5701504-5701504. Art no. 5701504. DOI: 10.1109/TASC.2011.2178984

[39] Zhang Z, et al. Characteristics of compensation for fluctuating output power of a solar power generator in a hybrid energy storage system using a Bi2223-SMES coil cooled by thermo-siphon with liquid hydrogen. In: IEEE Transactions on Applied Superconductivity. June 2016;**8223**(c):1-5. Art no. 5701005, DOI: 10.1109/TASC.2016.2529565

[40] Ngamroo I, Karaipoom T. Cooperative control of SFCL and SMES for enhancing fault ride through capability and smoothing Power fluctuation of DFIG Wind Farm. In: IEEE Transactions on Applied Superconductivity. Oct 2014;**24**(5):1-4. Art no. 5400304. DOI: 10.1109/TASC.2014.2340445

[41] Li J, Yang Q, Robinson F, Liang F, Zhang M, Yuan W. Design and test of a new droop control algorithm for a SMES/battery hybrid energy storage system. Energy. 2017;**118**:1110-1122. ISSN 0360-5442. DOI: 10.1016/j.energy.2016.10.130

[42] Zhu J, Bao X, Yang B, Chen P, Yang Y, Qiu M. Dynamic simulation test research on power fluctuation compensation using hybrid SMES of YBCO and BSCCO tapes. In: IEEE Transactions

on Applied Superconductivity. June 2012;**22**(3):5700404-5700404. Art no. 5700404. DOI: 10.1109/TASC.2011.2176091

[43] Mukherjee P, Rao VV. Effective location of SMES for power fluctuation mitigation of grid connected doubly fed induction generator. Journal of Energy Storage. 2020;**29**:101369

[44] Miyagi D, Sato R, Ishida N, Sato Y, Tsuda M, Hamajima T. Experimental research on compensation for power fluctuation of the renewable energy using the SMES under the state-of-current feedback control. June 2015;**25**(3):1-5. Art no. 5700305, DOI: 10.1109/TASC.2014.2368051

[45] Song M, Shi J, Liu Y, Xu Y, Hu N, Tang Y, et al. 100 kJ/50 kW HTS SMES for micro-grid. June 2015;**25**(3):1-6. 2015, Art no. 5700506, DOI: 10.1109/TASC.2014.2386345

[46] Yao W, Jiang L, Fang J, Wen J, Cheng S, Wu QH. Adaptive power oscillation damping controller of superconducting magnetic energy storage device for interarea oscillations in power system. International Journal of Electrical Power & Energy Systems. 2016;**78**:555-562

[47] Ngamroo I, Cuk Supriyadi AN, Dechanupaprittha S, Mitani Y. Power oscillation suppression by robust SMES in power system with large wind power penetration. Physica C: Superconductivity and its Applications. 2009;**469**(1):44-51

[48] Ngamroo I. Robust SMES controller design for stabilization of inter-area oscillation considering coil size and system uncertainties. Physica C: Superconductivity and its Applications. 2010;**470**(22):1986-1993

[49] Du W, Wang HF, Cheng S, Wen JY, Dunn R. Robustness of damping control implemented by energy storage systems installed in power systems. International Journal of Electrical Power & Energy Systems. 2011;**33**(1):35-42

[50] Xu Y, Ren L, Zhang Z, Tang Y, Shi J, Xu C, et al. Analysis of the loss and thermal characteristics of a SMES (superconducting magnetic energy storage) magnet with three practical operating conditions. Energy. 2018;**143**:372-384. ISSN 0360-5442. DOI: 10.1016/j.energy.2017.10.087

[51] Pahasa J, Ngamroo I. A heuristic training-based least squares support vector machines for power system stabilization by SMES. Expert Systems with Applications. 2011;**38**(11):13987-13993

[52] Saejia M, Ngamroo I. A robust centralized SMES controller design based on WAMS considering system and communication delay uncertainties. Electric Power Systems Research. 2011;**81**(4):846-852

[53] Rahim AHMA, Nowicki EP. A robust damping controller for an HV-ACDC system using a loop-shaping procedure. Journal of Electrical Engineering. 2005;**56**(1-2):15-20

[54] Tan YL, Wang Y. A robust nonlinear excitation and SMES controller for transient stabilization. International Journal of Electrical Power & Energy Systems. 2004;**26**(5):325-332

[55] Wang Y, Feng G, Cheng D, Liu Y. Adaptive L2 disturbance attenuation control of multi-machine power systems with SMES units. Automatica. 2006;**42**(7):1121-1132

[56] Ngamroo I. An optimization of robust SMES with specified structure H controller for power system stabilization considering superconducting magnetic coil size. Energy Conversion and Management. 2011;**52**(1):648-651

[57] Evans A, Strezov V, Evans TJ. Assessment of utility energy storage options for increased renewable energy penetration. Renewable and Sustainable Energy Reviews. 2012;**16**(6):4141-4147

[58] Ngamroo I, Vachirasricirikul S. Coordinated control of optimized SFCL and SMES for improvement of power system transient stability. In: IEEE Transactions on Applied Superconductivity. June 2012;**22**(3):5600805-5600805. Art no. 5600805. DOI: 10.1109/TASC.2011.2174550

[59] Peng J, Sun Y, Wang HF. Co-ordinated emergency control of generator-tripping and SMES based on Hamiltonian system theory. International Journal of Electrical Power & Energy Systems. 2005;**27**(5-6):352-360

[60] Kang BK, Kim ST, Bae SH, Park JW. Effect of a SMES in power distribution network with PV system and PBEVs. IEEE Transactions on Applied Superconductivity. 2013;**23**(3):3-6

[61] Noori A, Shahbazadeh MJ, Eslami M. Electrical power and energy systems designing of wide-area damping controller for stability improvement in a large-scale power system in presence of wind farms and SMES compensator. Electrical Power and Energy Systems. 2020;**119**:105936

[62] Antony AP, Shaw DT. Empowering the electric grid: Can SMES coupled to wind turbines improve grid stability? Renewable Energy. 2016;**89**:224-230

[63] Shi J, Tang Y, Xia Y, Ren L, Li J, Jiao F. Energy function based SMES controller for Transient Stability Enhancement In: IEEE Transactions on Applied Superconductivity. June 2012;**22**(3):5701304-5701304. Art no. 5701304. DOI: 10.1109/TASC.2011.2177431

[64] Ali MH, Park M, Yu IK, Murata T, Tamura J, Wu B. Enhancement of transient stability by fuzzy logic-controlled SMES considering communication delay. International Journal of Electrical Power & Energy Systems. 2009;**31**(7-8):402-408

[65] Fang J et al. Laboratory and field tests of movable conduction-cooled high-temperature SMES for power system stability enhancement. In: IEEE Transactions on Applied Superconductivity. 2013;**23**(4):5701607-5701607. Aug. 2013, Art no. 5701607. DOI: 10.1109/TASC.2013.2256350

[66] Yang J, Liu W, Liu P. Application of SMES unit in black start. Physics Procedia. 2014;**58**:277-281

[67] Khanna R, Singh G, Nagsarkar TK, Member S. Power System Stability Enhancement with SMES. International Conference on Power, Signals, Controls and Computation (EPSCICON). 3-6 Jan 2012:1-6

[68] Fang J, Yao W, Wen J, Cheng S, Tang Y, Cheng Z. Probabilistic assessment of power system transient stability incorporating SMES. Physica C: Superconductivity. 2013;**484**:276-281

[69] Tang Y, Mu C, He H. SMES-based damping controller design using fuzzy-GrHDP considering transmission delay. In: IEEE Transactions on Applied Superconductivity. Oct 2016;**26**(7)1-6. Art no. 5701206, DOI: 10.1109/TASC.2016.2586888

[70] Muyeen SM, Hasanien HM, Al-Durra A. Transient stability enhancement of wind farms connected to a multi-machine power system by using an adaptive ANN-controlled SMES.

Energy Conversion and Management. 2014;**78**:412-420

[71] Sadeghzadeh SM, Ehsan M, Hadj Said N, Feuillet R. Transient stability improvement of multi-machine power systems using on-line fuzzy control of SMES. Control Engineering Practice. 1999;**7**(4):531-536

[72] Rabbani MG, Devotta JBX, Elangovan S. A fuzzy set theory based control of superconductive magnetic energy storage unit to improve power system dynamic performance. Electric Power Systems Research. 1997;**40**(2):107-114

[73] Rabbani MG, Devotta JBX, Elangovan S. An efficient fuzzy controlled system for superconducting magnetic energy storage unit. International Journal of Electrical Power & Energy Systems. 1998;**20**(3):197-202

[74] Devotta JBX, Rabbani MG. Application of {superconducting magnetic energy storage} unit in multi-machine power systems. Energy Conversion and Management. 2000;**41**(5):493-504

[75] Yunus AMS, Abu-Siada A, Masoum MAS. Application of SMES unit to improve DFIG power dispatch and dynamic performance during intermittent misfire and fire-through faults. In: IEEE Transactions on Applied Superconductivity. Aug 2013;**23**(4):5701712-5701712. Art no. 5701712, DOI: 10.1109/TASC.2013.2256352

[76] Shi J, Tang Y, Dai T, Ren L, Li J, Cheng S. Determination of SMES capacity to enhance the dynamic stability of power system. Physica C: Superconductivity and its Applications. 2010;**470**(20):1707-1710

[77] Kopylov S, Balashov N, Ivanov S, Veselovsky A, Zhemerikin V. Use of superconducting devices operating together to ensure the dynamic stability of electric power system. IEEE Transactions on Applied Superconductivity. 2011;**21**(3 PART 2):2135-2139

[78] Abu-Siada A, Islam S. Application of SMES unit in improving the performance of an AC/DC power system. IEEE Transactions on Sustainable Energy. 2011;**2**(2):109-121

[79] Ali MH, Wu B, Tamura J, Dougal RA. Minimization of shaft oscillations by fuzzy controlled SMES considering time delay. Electric Power Systems Research. 2010;**80**(7):770-777

[80] Kim AR, Jung HY, Kim JH, Ali MH, Park M, Yu IK, et al. A study on the operation analysis of the power conditioning system with real HTS SMES coil. Physica C: Superconductivity and its Applications. 2008;**468**(15-20):2104-2110

[81] Molina MG, Mercado PE, Watanabe EH. Analysis of integrated STATCOM-SMES based on three-phase three-level multi-pulse voltage source inverter for high power utility applications. Journal of the Franklin Institute. 2011;**348**(9):2350-2377

[82] Shi J, Tang YJ, Ren L, Li JD, Chen SJ. Application of SMES in wind farm to improve voltage stability. Physica C: Superconductivity and its Applications. 2008;**468**(15-20):2100-2103

[83] Yunus AMS, Masoum MAS, Abu-Siada A. Application of SMES to enhance the dynamic performance of DFIG during voltage sag and swell. IEEE Transactions on Applied Superconductivity. Aug 2012;**22**(4):5702009-5702009.

Art no. 5702009. DOI: 10.1109/TASC.2012.2191769

[84] Said SM, Aly MM, Abdel-Akher M. Application of superconducting magnetic energy storage (SMES) for voltage sag/swell supression in distribution system with wind power penetration. In: 2014 16th International Conference on Harmonics and Quality of Power (ICHQP), Bucharest, Romania; 2014. pp. 92-96. DOI: 10.1109/ICHQP.2014.6842877

[85] Salama HS. Comparison of different electric vehicle integration approaches in presence of photovoltaic and superconducting magnetic energy storage systems. Journal of Cleaner Production. 2020;**260**:121099. ISSN 0959-6526. DOI: 10.1016/j.jclepro.2020.121099

[86] Chen XY, et al. Energy exchange experiments and performance evaluations using an equivalent method for a SMES prototype. In: IEEE Transactions on Applied Superconductivity. Oct 2014;**24**(5):1-5. Art no. 5701005. DOI: 10.1109/TASC.2014.2344759

[87] Zhu J, Cheng Q, Yang B, Yuan W, Coombs TA, Qiu M. Experimental research on dynamic voltage sag compensation using 2G HTS SMES. IEEE Transactions on Applied Superconductivity. 2011;**21**(3 PART 2):2126-2130

[88] A. Elnozahy and M. Elgamal, Minimum power loss based design of SMES as influenced by coil material, Journal of Energy Storage, vol. 30, no. December 2019, p. 101461, 2020.

[89] Huang XHX, Zhang GZG, Xiao LXL. Optimal location of SMES for improving power system voltage stability. IEEE Transactions on Applied Superconductivity. 2010;**20**(3):1316-1319

[90] Zheng ZX, Xiao XY, Li CS, Chen Z, Zhang Y. Performance evaluation of SMES system for initial and steady voltage sag compensations. In: IEEE Transactions on Applied Superconductivity. Oct 2016;**26**(7):1-5. Art no. 5701105, DOI: 10.1109/TASC.2016.2582844

[91] Shi J, Tang Y, Yang K, Chen L, Ren L, Li J, et al. SMES based dynamic voltage restorer for voltage fluctuations compensation. IEEE Transactions on Applied Superconductivity. 2010;**20**(3):1360-1364

[92] Kadam PS, Vadirajacharya K. Super Conducting Magnetic Energy Storage Unit for Power Conditioning. First International Conference on Advances in Computer, Electronics and Electrical Engineering - CEEE, Mumbai, India; 2012. DOI: 10.15224/978-981-07-1847-3-1018

[93] Shi J, Zhou A, Liu Y, Ren L, Tang Y, Li J. Voltage distribution characteristic of HTS SMES magnet. In: IEEE Transactions on Applied Superconductivity. June 2016;**26**(4):1-5. Art no. 5700705, DOI: 10.1109/TASC.2016.2536656

[94] Farahani M. A new control strategy of SMES for mitigating subsynchronous oscillations. Physica C: Superconductivity and its Applications. 2012;**483**:34-39

[95] Gil-gonzález W, Danilo O, Garces A. Control of a SMES for mitigating subsynchronous oscillations in power systems: A PBC-PI approach. Journal of Energy Storage. 2018;**20**(September):163-172

[96] Sedighizadeh M, Esmaili M, Parvaneh H. Coordinated optimization

and control of SFCL and SMES for mitigation of SSR using HBB-BC algorithm in a fuzzy framework. Journal of Energy Storage. 2018;**18**(January):498-508

[97] Abu-Siada A, Abu-Siada A, Pota HR. Damping of subsynchronous oscillations and improve transient stability for wind farms. 2011 IEEE PES Innovative Smart Grid Technologies, Perth, WA, Australia; 2011. pp. 1-6. DOI: 10.1109/ISGT-Asia.2011.6167077

[98] Rabbani MG, Elangovan S. Multi-mode wide range subsynchronous resonance stabilization using superconducting magnetic energy storage unit. International Journal of Electrical Power & Energy Systems. 1999;**21**:45-53. ISSN 0142-0615. DOI: 10.1016/S0142-0615(98)00029-5

[99] Aly MM, Abdel-Akher M, Said SM, Senjyu T. A developed control strategy for mitigating wind power generation transients using superconducting magnetic energy storage with reactive power support. International Journal of Electrical Power & Energy Systems. 2016;**83**:485-494

[100] Shivarama Krishna K, Sathish Kumar K. A review on hybrid renewable energy systems. Renewable and Sustainable Energy Reviews. 2015;**52**:907-916

[101] Hasanien HM. A set-membership affine projection algorithm-based adaptive-controlled SMES units for wind farms output power smoothing. IEEE Transactions on Sustainable Energy. 2014;**5**(4):1226-1233

[102] Hamajima T, Tsuda M, Miyagi D, Amata H, Iwasaki T, Son K, et al. Advanced superconducting power conditioning system with SMES for effective use of renewable energy. Physics Procedia. 2012;**27**:396-399

[103] Li J, Wang X, Zhang Z, Le S, Yang Q, Zhang M, et al. Analysis of a new design of the hybrid energy storage system used in the residential m-CHP systems. Applied Energy. 2017;**187**:169-179

[104] Li J, Gee AM, Zhang M, Yuan W. Analysis of battery lifetime extension in a SMES-battery hybrid energy storage system using a novel battery lifetime model. Energy. 2015;**86**:175-185

[105] Liu Y, Tang Y, Shi J, Shi X, Deng J, Gong K. Application of small-sized SMES in an EV charging station with DC bus and PV system. In: IEEE Transactions on Applied Superconductivity. June 2015;**25**(3):1-6. Art no. 5700406. DOI: 10.1109/TASC.2014.2374174

[106] Hamajima T, Amata H, Iwasaki T, Atomura N, Tsuda M, Miyagi D, et al. Application of SMES and fuel cell system combined with liquid hydrogen vehicle station to renewable energy control. IEEE Transactions on Applied Superconductivity. 2012;**22**(3):3-6

[107] Kim ST, Kang BK, Bae SH, Park JW. Application of SMES and grid code compliance to wind/photovoltaic generation system. In: IEEE Transactions on Applied Superconductivity. 2012;**23**(3):5000804-5000804. June 2013, Art no. 5000804. DOI: 10.1109/TASC.2012.2232962

[108] Gong K, Shi J, Liu Y, Wang Z, Ren L, Zhang Y. Application of SMES in the micro-grid based on fuzzy control. In: IEEE Transactions on Applied Superconductivity. April 2016;**26**(3):1-5. Art no. 3800205. DOI: 10.1109/TASC.2016.2524446

[109] Shim JW, Cho Y, Kim S, Min SW, Hur K. Comments and corrections

corrections to synergistic control of SMES and battery energy storage for enabling dispatchability of renewable energy sources. In: IEEE Transactions on Applied Superconductivity. Dec 2014;**24**(6):9700101. Art no. 9700101. DOI: 10.1109/TASC.2014.2349391

[110] Yang B, Wang J, Zhang X, Yu L, Shu H, Yu T. Control of SMES systems in distribution networks with renewable energy integration: A perturbation estimation approach. Energy. 2020;**202**:117753

[111] Bhatt P, Ghoshal SP, Roy R. Coordinated control of TCPS and SMES for frequency regulation of interconnected restructured power systems with dynamic participation from DFIG based wind farm. Renewable Energy. 2012;**40**(1):40-50

[112] Wang Z, Zou Z, Zheng Y. Design and control of a photovoltaic energy and SMES hybrid system with current-source grid inverter. In: IEEE Transactions on Applied Superconductivity. June 2013;**23**(3):5701505-5701505. Art no. 5701505. DOI: 10.1109/TASC.2013.2250172

[113] Sun Q, Xing D, Alafnan H, Pei X, Zhang M, Yuan W. Design and test of a new two-stage control scheme for SMES-battery hybrid energy storage systems for microgrid applications, Applied Energy, vol. 253, no. September 2018. 2019. p. 113529

[114] Gouda EA, Abd-alaziz A, El-saadawi M. Design modeling, and control of multi-stage SMES integrated with PV system. Journal of Energy Storage. 2020;**29**:101399

[115] Ngamroo I, Vachirasricirikul S. Design of optimal SMES controller considering SOC and robustness for microgrid stabilization. In IEEE Transactions on Applied Superconductivity. Oct 2016;**26**(7):1-5. Art no. 5403005, DOI: 10.1109/TASC.2016.2597261

[116] Shintomi T, et al. Design study of MgB_2 SMES coil for effective use of renewable energy. In: IEEE Transactions on Applied Superconductivity. June 2013;**23**(3):5700304-5700304. Art no. 5700304. DOI: 10.1109/TASC.2012.2234181

[117] Shintomi T, et al. Design study of SMES system cooled by thermo-siphon with liquid hydrogen for effective use of renewable energy. In: IEEE Transactions on Applied Superconductivity. June 2012;**22**(3):5701604-5701604. Art no. 5701604. DOI: 10.1109/TASC.2011.2178575

[118] Tixador P. Development of superconducting power devices in Europe. Physica C: Superconductivity and its Applications. 2010;**470**(20):971-979

[119] Bizon N. Effective mitigation of the load pulses by controlling the battery/SMES hybrid energy storage system. Applied Energy. 2018;**229**(August):459-473

[120] Koohi-Kamali S, Tyagi VV, Rahim NA, Panwar NL, Mokhlis H. Emergence of energy storage technologies as the solution for reliable operation of smart power systems: A review. Renewable and Sustainable Energy Reviews. 2013;**25**:135-165

[121] Hemmati R, Saboori H. Emergence of hybrid energy storage systems in renewable energy and transport applications – A review. Renewable and Sustainable Energy Reviews. 2016;**65**:11-23

[122] Yekini Suberu M, Wazir Mustafa M, Bashir N. Energy storage systems for renewable energy power sector integration and mitigation

of intermittency. Renewable and Sustainable Energy Reviews. 2014;**35**:499-514

[123] Vazquez S, Lukic SM, Galvan E, Franquelo LG, Carrasco JM. Energy storage systems for transport and grid applications. IEEE Transactions on Industrial Electronics. 2010;**57**(12):3881-3895

[124] Zhu J, Yuan W, Qiu M, Wei B, Zhang H, Chen P, et al. Experimental demonstration and application planning of high temperature superconducting energy storage system for renewable power grids. Applied Energy. 2015;**137**:692-698

[125] Molina MG, Mercado PE, Hirokazu Watanabe E. Improved superconducting magnetic energy storage (SMES) controller for high-power utility applications. IEEE Transactions on Energy Conversion. 2011;**26**(2):444-456

[126] Wang Z, Yuwen B, Lang Y, Cheng M. Improvement of operating performance for the wind farm with a novel CSC-type wind turbine-SMES hybrid system. 2012 IEEE International Symposium on Industrial Electronics. Hangzhou, China; 2012:1017-1022. DOI: 10.1109/ISIE.2012.6237228

[127] Shi J, Xu Y, Liao M, Guo S, Li Y, Ren L, et al. Integrated design method for superconducting magnetic energy storage considering the high frequency pulse width modulation pulse voltage on magnet. Applied Energy. 2019;**248**(January):1-17

[128] Chen XY, et al. Integrated SMES technology for modern power system and future smart grid. In: IEEE Transactions on Applied Superconductivity. Oct 2014;**24**(5):1-5. Art no. 3801606. DOI: 10.1109/TASC.2014.2346502

[129] Kang J, Member S, Ko TK. Jointless pancake coil winding for minimizing electrical loss in HTS SMES for wind Power. In: IEEE Transactions on Applied Superconductivity. June 2015;**25**(3):1-5. Art no. 5700705. DOI: 10.1109/TASC.2015.2390620

[130] Colmenar-santos A, Molina-ibáñez E, Rosales-asensio E, Blanes-peiró J. Legislative and economic aspects for the inclusion of energy reserve by a superconducting magnetic energy storage: Application to the case of the Spanish electrical system. Renewable and Sustainable Energy Reviews. 2017:1-16

[131] Sander M, Gehring R, Neumann H. LIQHYSMES — A 48 GJ toroidal MgB2-SMES for buffering minute and second fluctuations. In: IEEE Transactions on Applied Superconductivity. June 2013;**23**(3):5700505-5700505. Art no. 5700505, DOI: 10.1109/TASC.2012.2234201

[132] Sander M, Brighenti F, Gehring R, Jordan T, Klaeser M, Kraft D, et al. LIQHYSMES - liquid H2 and SMES for renewable energy applications. International Journal of Hydrogen Energy. 2014;**39**(23):12007-12017

[133] Sander M, Gehring R, Neumann H, Jordan T. LIQHYSMES storage unit - hybrid energy storage concept combining liquefied hydrogen with superconducting magnetic energy storage. International Journal of Hydrogen Energy. 2012;**37**(19):14300-14306

[134] Sander M, Gehring R. LIQHYSMES—A novel energy storage concept for variable renewable energy sources using hydrogen and SMES. IEEE Transactions on Applied Superconductivity. 2011;**21**(3):1362-1366

[135] Ayodele TR, Ogunjuyigbe ASO. Mitigation of wind power intermittency:

Storage technology approach. Renewable and Sustainable Energy Reviews. 2015;**44**:447-456

[136] Wang Z, Jiang L, Zou Z, Cheng M. Operation of SMES for the current source inverter fed distributed power system under islanding mode. In: IEEE Transactions on Applied Superconductivity. June 2013;**23**(3):5700404-5700404. Art no. 5700404. DOI: 10.1109/TASC.2012.2232703

[137] Fathima AH, Palanisamy K. Optimization in microgrids with hybrid energy systems - a review. Renewable and Sustainable Energy Reviews. 2015;**45**:431-446

[138] M. H. Qais, H. M. Hasanien, and S. Alghuwainem, Output power smoothing of wind power plants using self-tuned controlled SMES units, Electric Power Systems Research, vol. 178, no. April 2019, p. 106056, 2020.

[139] Gil-gonzález W, Danilo O. Passivity-based PI control of a SMES system to support power in electrical grids: A bilinear approach. Journal of Energy Storage. 2018;**18**(May):459-466

[140] Akinyele DO, Rayudu RK. Review of energy storage technologies for sustainable power networks. Sustainable Energy Technologies and Assessments. 2014;**8**:74-91

[141] Lund PD, Lindgren J, Mikkola J, Salpakari J. Review of energy system flexibility measures to enable high levels of variable renewable electricity. Renewable and Sustainable Energy Reviews. 2015;**45**:785-807

[142] Ngamroo I. Robust SMES controller design based on inverse additive perturbation for stabilization of interconnected power systems with wind farms. Energy Conversion and Management. 2010;**51**(3):459-464

[143] Xie Y, Song M, Shi J, Jiang G, Geng P, Zhang M. Simulation on a micro-grid system based on superconducting magnetic energy storage. In: 2014 International Power Electronics and Application Conference and Exposition. IEEE PEAC 2014. Shanghai, China; 2014. pp. 1451-1455. DOI: 10.1109/PEAC.2014.7038078

[144] Georgescu M, Barote L, Marinescu C, Clotea L. Smart electrical energy storage system for small power wind turbines. In: 2010 12th International Conference on Optimization of Electrical and Electronic Equipment. 2010. pp. 1192-1197. DOI: 10.1109/OPTIM.2010.5510524

[145] Gao S, Chau KT, Liu C, Wu D, Li J. SMES control for power grid integrating renewable generation and electric vehicles. IEEE Transactions on Applied Superconductivity. 2012;**22**(3):3-6

[146] Nie Z, Xiao X, Kang Q, Aggarwal R, Zhang H, Yuan W. SMES-battery energy storage system for conditioning outputs from direct drive. Linear Wave Energy Converters. In: IEEE Transactions on Applied Superconductivity. June 2013;**23**(3):5000705-5000705. Art no. 5000705. DOI: 10.1109/TASC.2013.2246852

[147] Saejia M, Ngamroo I. Stabilization of microgrid with intermittent renewable energy sources by SMES with optimal coil size. Physica C: Superconductivity and its Applications. 2011;**471**(21-22):1385-1389

[148] Liu Y et al. Status evaluation method for SMES used in power grid. In: IEEE Transactions on Applied Superconductivity. Oct 2015;**25**(5):1-10. Art no. 5701310, DOI: 10.1109/TASC.2015.2456106

[149] Nielsen KE, Molinas M. Superconducting magnetic energy storage (SMES) in power systems with renewable energy sources. In: 2010 IEEE International Symposium on Industrial Electronics, Bari, Italy. 2010. pp. 2487-2492. DOI: 10.1109/ISIE.2010.5637892

[150] Shim JW, Cho Y, Kim S, Min SW, Hur K, Member S. Synergistic control of SMES and battery energy storage for enabling dispatchability of renewable energy sources. In: IEEE Transactions on Applied Superconductivity. June 2013;**23**(3):5701205-5701205. Art no. 5701205, DOI: 10.1109/TASC.2013.2241385

[151] Nam T, Shim JW, Hur K. The beneficial role of SMES coil in DC lines as an energy buffer for integrating large scale wind power. In: IEEE Transactions on Applied Superconductivity. June 2012;**22**(3):5701404-5701404. Art no. 5701404. DOI: 10.1109/TASC.2011.2175686

[152] Tam KS, Kumar P, Foreman M. Using SMES to support large-scale PV power generation. Solar Energy. 1990;**45**(1):35-42

[153] Kumar NJV, Thameem Ansari MM. A new design of dual-mode type-II fuzzy logic load frequency controller for interconnected power systems with parallel AC-DC tie-lines and superconducting magnetic energy storage unit. Energy. 2015;**89**:118-137

[154] Shayeghi H, Jalili A, Shayanfar HA. A robust mixed H2/H1 based LFC of a deregulated power system including SMES. Energy Conversion and Management. 2008;**49**(10):2656-2668

[155] Farhadi Kangarlu M, Alizadeh Pahlavani MR. Cascaded multilevel converter based superconducting magnetic energy storage system for frequency control. Energy. 2014;**70**:504-513

[156] Li J, Xiong R, Yang Q, Liang F, Zhang M, Yuan W. Design/test of a hybrid energy storage system for primary frequency control using a dynamic droop method in an isolated microgrid power system. Applied Energy. 2017;**201**:257-269

[157] Zhu J, Qiu M, Wei B, Zhang H, Lai X, Yuan W. Design, dynamic simulation and construction of a hybrid HTS SMES (high-temperature superconducting magnetic energy storage systems) for Chinese power grid. Energy. 2013;**51**:184-192

[158] Nandi M, Shiva CK, Mukherjee V. Frequency stabilization of multi-area multi-source interconnected power system using TCSC and SMES mechanism. Journal of Energy Storage. 2017;**14**(2):348-362. ISSN 2352-152X, DOI: 10.1016/j.est.2017.10.018

[159] Selvaraju RK, Somaskandan G. Impact of energy storage units on load frequency control of deregulated power systems. Energy. 2016;**97**:214-228

[160] Pappachen A, Fathima AP. Load frequency control in deregulated power system integrated with SMES-TCPS combination using ANFIS controller. International Journal of Electrical Power & Energy Systems. 2016;**82**:519-534

[161] Sudha KR, Vijaya Santhi R. Load frequency control of an interconnected reheat thermal system using Type-2 fuzzy system including SMES units. International Journal of Electrical Power & Energy Systems. 2012;**43**(1):1383-1392

[162] Ganapathy S, Velusami S. MOEA based design of decentralized controllers for LFC of interconnected power systems with nonlinearities, AC-DC parallel tie-lines and SMES units. Energy Conversion and Management. 2010;**51**(5):873-880

[163] Elsisi M, Soliman M, Aboelela MAS, Mansour W. Optimal design of model predictive control with superconducting magnetic energy storage for load frequency control of nonlinear hydrothermal power system using bat inspired algorithm. Journal of Energy Storage. 2017;**12**:311-318

[164] Kim AR, Kim SY, Kim KM, Kim JG, Kim S, Park M, et al. Performance analysis of a toroid-type HTS SMES adopted for frequency stabilization. IEEE Transactions on Applied Superconductivity. 2011;**21**(3 PART 2):1367-1370

[165] Kim AR, Kim GH, Heo S, Park M, Yu IK, Kim HM. SMES application for frequency control during islanded microgrid operation. Physica C: Superconductivity and its Applications. 2013;**484**:282-286

[166] Farahani M, Ganjefar S. Solving LFC problem in an interconnected power system using superconducting magnetic energy storage. Physica C: Superconductivity and its Applications. 2013;**487**:60-66

[167] Demiroren A, Yesil E. Automatic generation control with fuzzy logic controllers in the power system including SMES units. International Journal of Electrical Power & Energy Systems. 2004;**26**(4):291-305

[168] Bhatt P, Roy R, Ghoshal SP. Comparative performance evaluation of SMES-SMES, TCPS-SMES and SSSC-SMES controllers in automatic generation control for a two-area hydro-hydro system. International Journal of Electrical Power & Energy Systems. 2011;**33**(10):1585-1597

[169] Chaine S, Tripathy M. Design of an optimal SMES for automatic generation control of two-area thermal power system using cuckoo search algorithm. Journal of Electrical Systems and Information Technology. 2015;**2**(1):1-13

[170] Pradhan PC, Sahu RK, Panda S. Firefly algorithm optimized fuzzy PID controller for AGC of multi-area multi-source power systems with UPFC and SMES. Engineering Science and Technology, an International Journal. 2015;**19**(1):338-354

[171] Zhang JY, et al. Electric energy exchange and applications of superconducting magnet in an SMES device. In: IEEE Transactions on Applied Superconductivity. June 2014;**24**(3): 1-4. Art no. 5700704. DOI: 10.1109/TASC.2013.2291438

[172] Ngamroo I, Karaipoom T. Improving low-voltage ride-through performance and alleviating power fluctuation of DFIG wind turbine in DC microgrid by optimal SMES, with fault current limiting function. In: IEEE Transactions on Applied Superconductivity. Oct 2014;**24**(5):1-5. Art no. 5700805. DOI: 10.1109/TASC.2014.2333031

[173] Sedighizadeh M, Yarmohammadi H, Esmaili M. Engineering science and technology, an International journal enhancing FRT performance and smoothing output power of DFIG wind farm equipped by SFCL and SMES in a fuzzy framework. Engineering Science and Technology, an International Journal. 2019;**22**(3):801-810

[174] Ngamroo I. Optimization of SMES-FCL for augmenting FRT performance and smoothing output power of Grid-Connected DFIG Wind Turbine. In: IEEE Transactions on Applied Superconductivity. Oct 2016;**26**(7): 1-5, Art no. 3800405, DOI: 10.1109/TASC.2016.2592945

[175] Jin JX, Chen XY. Cooperative operation of superconducting

fault-current-limiting cable and SMES system for grounding fault protection in a LVDC network. IEEE Transactions on Industry Applications. 2015;**51**(6):5410-5414

[176] Ngamroo I, Vachirasricirikul S. Optimized SFCL and SMES units for multimachine transient stabilization based on kinetic energy control. In: IEEE Transactions on Applied Superconductivity. June 2013;**23**(3):5000309-5000309. Art no. 5000309. DOI: 10.1109/TASC.2013.2240760

[177] Mohamed EA, Gouda E, Mitani Y. Impact of SMES integration on the digital frequency relay operation considering High PV/Wind penetration in micro-grid, Energy Procedia. 2019;**157**:1292-1304. ISSN 1876-6102. DOI: 10.1016/j.egypro.2018.11.295

[178] Sekhar R, Kumar A, Rajesh G. Superconducting magnetic energy storage (SMES) devices integrated with resistive type superconducting fault current limiter (SFCL) for fast recovery time. Journal of Energy Storage. 2017;**13**:287-295

[179] Nomura S, Chikaraichi H. Feasibility study on large scale SMES for daily load leveling using force-balanced helical coils. In: IEEE Transactions on Applied Superconductivity. June 2013;**23**(3):5700904-5700904. Art no. 5700904. DOI: 10.1109/TASC.2012.2237494

[180] Saranya S, Saravanan B. Optimal size allocation of superconducting magnetic energy storage system based unit commitment. Journal of Energy Storage. 2018;**20**(May):173-189

[181] Molina MG, Mercado PE. Primary frequency control of multi-machine power systems with STATCOM-SMES: A case study. International Journal of Electrical Power & Energy Systems. 2013;**44**(1):388-402

[182] Xing YQ, Jin JX, Wang YL, Du BX, Wang SC, Modeling AS. An electric vehicle charging System using an SMES. Implanted Smart Grid. In: IEEE Transactions on Applied Superconductivity. Oct 2016;**26**(7):1-4. Art no. 5701504. DOI: 10.1109/TASC.2016.2602245

[183] Yang B, Zhu T, Zhang X, Wang J, Shu H, Li S, et al. Design and implementation of battery/SMES hybrid energy storage systems used in electric vehicles: A nonlinear robust fractional-order control approach. Energy. 2020;**191**:116510. ISSN 0360-5442. DOI: 10.1016/j.energy.2019.116510

[184] Ren G, Ma G, Cong N. Review of electrical energy storage system for vehicular applications. Renewable and Sustainable Energy Reviews. 2015;**41**:225-236

[185] González-Gil A, Palacin R, Batty P. Sustainable urban rail systems: Strategies and technologies for optimal management of regenerative braking energy. Energy Conversion and Management. 2013;**75**:374-388

[186] Rassõlkin A, Hõimoja H. Switching locomotive as a part of smart electrical grid. IFAC Proceedings Volumes. 2012;**8**(PART 1):606-609

[187] Wu D, Chau KT, Liu C, Gao S, Li F. Transient stability analysis of SMES for smart grid with vehicle-to-grid operation. In: IEEE Transactions on Applied Superconductivity. June 2012;**22**(3):5701105-5701105. Art no. 5701105. DOI: 10.1109/TASC.2011.2174572

[188] Kazi Shariful I, Mehdi S, Alex S. Decentralized robust mixed H2/H∞

reactive power control of DFIG cluster using SMES, International Journal of Electrical Power & Energy Systems. 2019;**113**:176-187. ISSN 0142-0615. DOI: 10.1016/j.ijepes.2019.05.017

[189] Gil-gonzález W, Danilo O. Active and reactive power conditioning using SMES devices with PMW-CSC: A feedback nonlinear control approach. Ain Shams Engineering Journal. 2019;**10**(2):369-378. ISSN 2090-4479. DOI: 10.1016/j.asej.2019.01.001

[190] Ananthavel S, Padmanaban S, Shanmugham S, Blaabjerg F, Ertas AH, Fedak V. Analysis of enhancement in available power transfer capacity by STATCOM integrated SMES by numerical simulation studies. Engineering Science and Technology, an International Journal. 2015;**19**(2):671-675. ISSN 2215-0986. DOI: 10.1016/j.jestch.2015.10.002

[191] Danilo O, Gil-gonzález W, Garcés A, Espinosa-pérez G. Indirect IDA-PBC for active and reactive power support in distribution networks using SMES systems with PWM-CSC. Journal of Energy Storage. 2018;**17**:261-271

[192] Lee Y. Superconducting magnetic energy storage controller design and stability analysis for a power system with various load characteristics. Electric Power Systems Research. 1999;**51**(1):33-41. ISSN 0378-7796. DOI: 10.1016/S0378-7796(98)00161-8

[193] Kumar A, Lal JVM, Agarwal A. Electromagnetic analysis on 2. 5MJ high temperature superconducting magnetic energy storage (SMES) coil to be used in uninterruptible power applications. Materials Today: Proceedings. 2020;**21**:1755-1762

Chapter 3

Battery-Assisted PV-Pumping System

Ahmed Alaa Mahfouz

Abstract

Photovoltaic (PV) water pumping systems convert solar radiation into electricity *via* PV panels to feed and drive electric pumps. The electrical energy is produced in DC form and must be adopted and then converted into alternating current by employing inverters. Batteries are used to store energy and to improve the system performance. Solar battery-assisted water pumping technology is on the long-run cost-effective and sure environment friendly. PV must be controlled to achieve the maximum power point operating condition. The state of charge of the battery should be observed to guarantee a safe and long battery working hours. In this chapter, a simple and efficient off-grid battery-assisted quasi Z-source inverter (QZSI) was developed in a PV-pumping system as a replacement to the traditional two-stage converter (boost converter + voltage source inverter). Analysis and mathematical model of QZSI has been presented. MPPT from the solar panels and the battery state of charge (SOC) are to be considered in determining controller's parameters. The controller produces the shoot through duty ratio (D) and modulation index of the inverter (M). The inverter switching pattern is prepared based on the simple boost modulation technique. The performance of the system under study is verified *via* MATLAB/Simulink simulation.

Keywords: quasi-Z source inverter, energy storage, PV-pumping, maximum power point tracking, simple boost modulation

1. Introduction

Integration of renewable energy sources into the electrical generation systems is badly needed because of environmental and economic considerations. Solar energy is, of course, one of the most promising alternatives. However, a solar energy panel suffers from its need to a large space, low overall efficiency, and high initial and erection costs. In order to get an acceptable overall system efficiency, the interfacing systems (i.e., voltage regulators, inverters, etc.) must attain a very high efficiency.

Standalone systems and grid connected systems are two kinds of solar energy harvesters. Standalone systems are used in areas and sites where the grid is not reachable. A standalone system needs to be assisted by an energy storage unit to ensure a continuous and a better power quality. On the other hand, there is no need to have an energy storage unit in grid connected systems. Moreover, the energy harvesters improve the grid capabilities. The grid connected systems, however, have their own complexity and cost represented by phase-locked loops (PLL) for synchronization, active ad reactive power control and extra protection circuits.

In 2003, the impedance source inverter (ZSI) was introduced as a single-stage inverter [1]. An impedance network is included separating the source and the main inverter bridge as shown in **Figure 1**. The inverter is controlled by the same PWM technique applied in traditional two-stage inverter. However, the DC link voltage can be boosted by utilizing the additional shoot-through (ST) states, where any or all of the inverter bridge legs can be shorted. The boosting factor is determined by the duration of the ST periods. A modified topology of Z-source inverters was introduced in 2008 namely; quasi Z-source inverter (QZSI). The new topology features, in addition to all the advantages of ZSI, smaller passive element requirements for the impedance network [1, 2].

In literature, four main quasi Z-source inverters have been developed which feature several improvements when compared to the traditional ZSIs. The voltage-fed ZSI as well as the two QZS voltage-fed inverters, with similar properties to the ZSI, are shown in **Figure 2**(**a**, **c**, and **e**), while the current-fed ZSI as well as two QZS current-fed inverters are shown in **Figure 2**(**b**, **d** and **f**).

The QZSI topologies shown in **Figure 2c** and **e**, are of unidirectional nature as the voltage-fed ZSI. In order to gain the bidirectional property, the diode, D1 could be substituted by a switch bidirectional in conducting, unidirectional in blocking. The capacitor C_2 in QZSI, shown in **Figure 2c** has a lower voltage value than the same capacitor C_2 in ZSI shown in **Figure 2a**. In the QZSI circuit, illustrated in **Figure 2e**, both capacitors C1 and C2 have lower voltages. Input current in circuit in **Figure 2c** is continuous and discontinuous in circuit in **Figure 2e**. There is no need to include an input capacitance in the topology of the QZSI shown in **Figure 2c**, due to presence the input inductor, L_1. Input capacitance is essential in both topologies of voltage-fed ZSI and QZSI in **Figure 2e**.

Circuits in **Figure 2d** and **f** are current-fed QZSI topologies. The presence of the diode, D1, with its specified position, in both circuits of **Figure 2d** and **f** marks these topologies with bidirectional property.

Current rating of inductor L_2 in the QZSI shown in **Figure 2d** is lower than the current rating of L_2 in the ZSI shown in **Figure 2b**. Reduced passive component count is clearly noticed. The QZSI shown in **Figure 2f** has also lower current ratings for inductors L_1 and L_2.

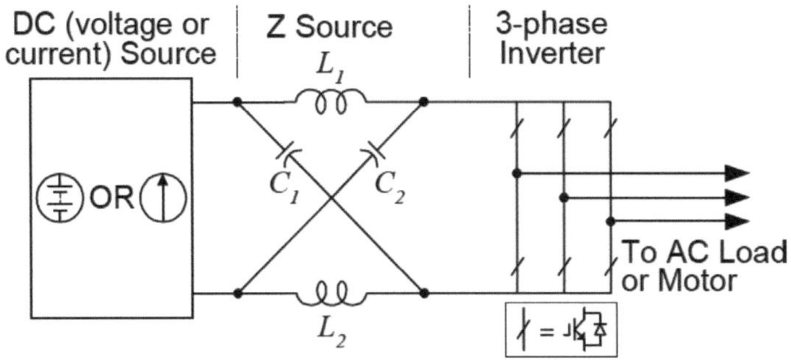

Figure 1.
Impedance source inverter.

DOI: http://dx.doi.org/10.5772/intechopen.110890

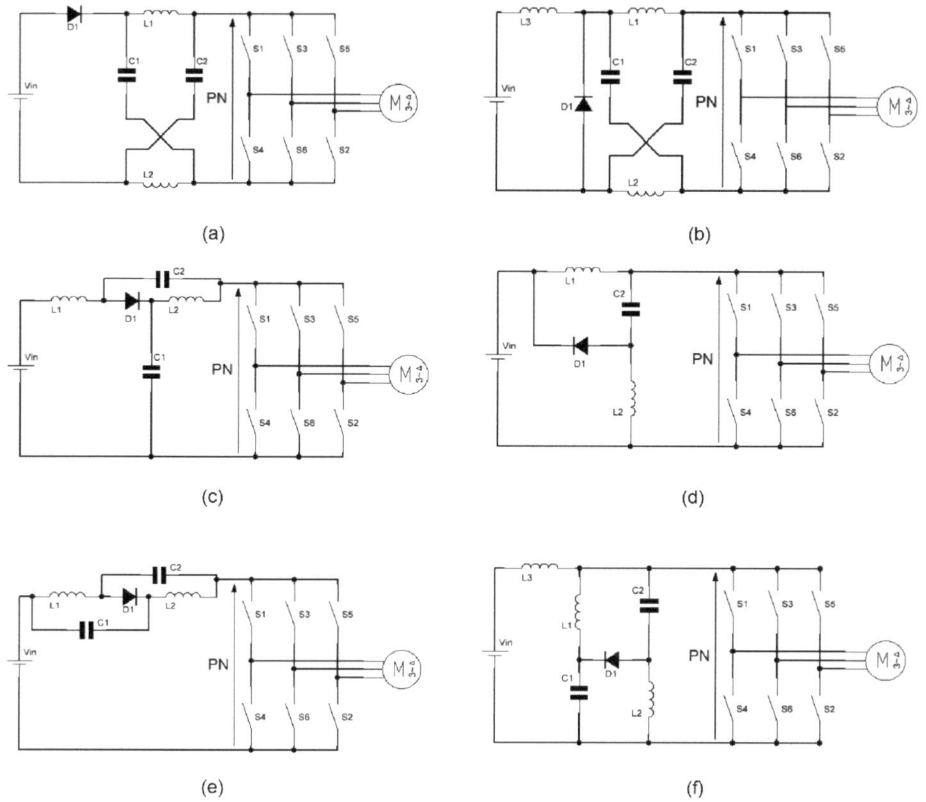

Figure 2.
Different topologies of Z-source and quasi Z-source inverters. a. Voltage-fed ZSI. b. Current-fed ZSI. c. Voltage-fed QZSI with continuous current. d. Current-fed QZSI with discontinuous current. e. Voltage-fed QZSI with discontinuous current. f. Current-fed QZSI with continuous current.

A common DC rail between the source and inverter characterizes the four QZSI voltage-fed and current-fed topologies. Additionally, the four QZSI circuits have no disadvantages in comparison with the conventional ZSI basic circuits. The four QZSI topologies can replace ZSI topology in any application [3].

In this chapter, a quasi Z-source inverter shown in **Figure 2c** will be adopted. These inverters are capable of performing maximum power tracking and inversion with no need for extra DC-DC converter. Other important operating characteristic is that the QZSI is operated with continuous input current. This ensures continuous supply current in PV system to facilitate maximization of the energy harvested. In addition, the system will be equipped with an energy storage device integrated into its topology as shown in **Figure 3**. This feature is essential when operating at low PV power conditions. In order to operate under these conditions, an energy storage device capable of managing the load demand for a period of time is required. Different placements of the batteries are used in literature as shown in **Figure 3**. By shunting a battery with either of the two circuit capacitors, the pumping system can be powered smoothly regardless the variations or the fluctuations of PV panel output. Because of the unique impedance network of QZSI, there is no need to add any extra circuit to charge or discharge the battery.

The control of the QZSI becomes more complex if the battery is added to the topology. For the sake of keeping the battery lasts longer, the accurate control of the battery

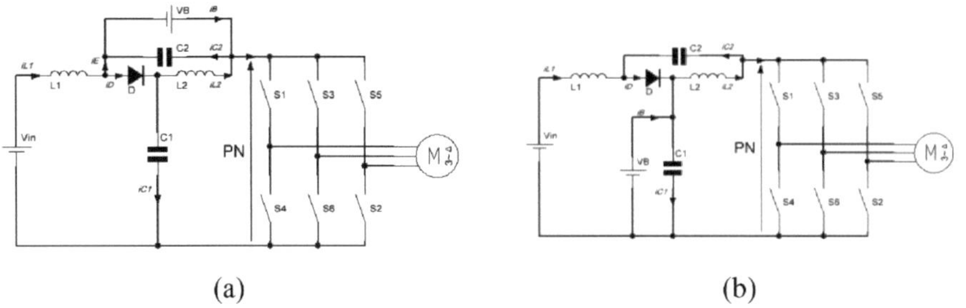

Figure 3.
Topology of battery-assisted QZSI. (a) Option 1 (b) option 2.

charging current should be carefully treated. However, many papers pay attention to other aspects, regarding energy (production and management), system (reliability, unit size and cost). In this research, battery charging control methodology for QZSI with energy storage will be presented along with controlling the output power of the inverter.

2. Analysis of quazi Z-source inverter

2.1 Battery-assisted QZSI

A topology improvement for the QZSI to be used in PV systems was introduced in [4, 5], where an energy storage device is added to the circuit without any extra circuitry depending on QZSI unique input impedance network. This is available also in Z-source inverters as shown in [6–8]. However, they experience discontinuous input current that is not desirable for PV system. Moreover, batteries suffer from higher voltage rating than that the one expected in the case of the ZSI. The QZSI topologies with storage units operate with continuous input current and so they have low voltage rating batteries.

2.1.1 Circuit analysis for battery parallel with C_2

The proposed QZSI with energy storage topology is shown in **Figure 3(a)**. These types of inverters are operated in two states, namely firstly, the active state which is pointed to as (T_1), and secondly, the shoot-through state which is pointed to as (T_0). In each switching cycle, the periodic time is calculated as $T = T_1 + T_0$. The inverter is operated based on normal sinusoidal pulse width modulation (SPWM) during the active state. While a short circuit between terminal P and terminal N is formed during the shoot-through state. The specified QZSI LC network is actually acting as a step-up DC-DC converter. Such converter is controlled by the shoot-through state. The equivalent circuit for each of the states is given in **Figures 4** and **5**. These equivalent circuits are accurate and valid as long as the current through each inductor is continuous so that the diode is complementary with the shoot-through state [9].

2.2 Shoot-through state mode

This mode (the shoot-through state) will make the inverter short circuit. As a result, the diode is turned off due to the reverse-bias voltage. The equivalent circuit,

DOI: http://dx.doi.org/10.5772/intechopen.110890

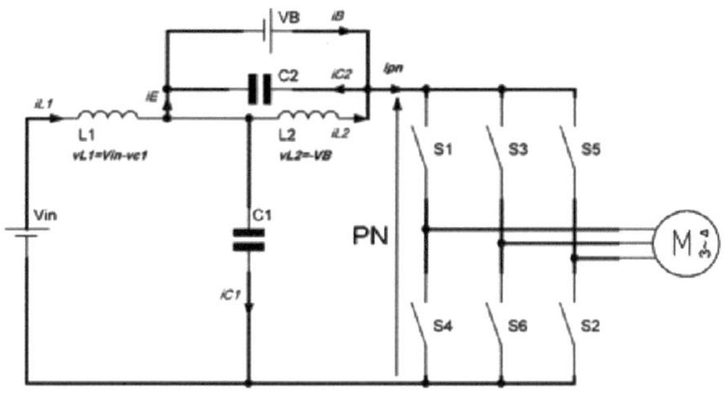

Figure 4.
Non-shoot-through state of battery-assisted QZSI [9].

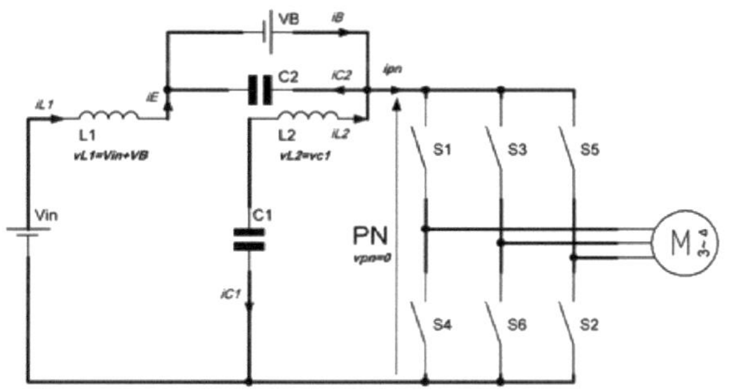

Figure 5.
Shoot-through state of battery-assisted QZSI [9].

representing this mode, is shown in **Figure 5**. During this time interval, the circuit equations are formed as follows:

$$C\frac{dV_{C1}}{dt} = i_B - i_{L2} \tag{1}$$

$$C\frac{dV_{C2}}{dt} = -i_{L1} \tag{2}$$

$$L\frac{di_{L1}}{dt} = V_{in} + V_{C2} \tag{3}$$

$$L\frac{di_{L2}}{dt} = V_{C1} \tag{4}$$

Where i_{L1}, i_{L2} and i_B are the currents in inductors L_1 and L_2, and in battery, respectively; V_{C1}, V_{C2} and V_{in} are the voltages across capacitors C_1, C_2 and PV panel, respectively; C is the capacitance of capacitors C_1 and C_2; L is the inductance of inductors L_1 and L_2.

2.3 Non-shoot-through state mode

In this mode, the inverter is free to follow any one of the six active (non-zero) states and two conventional zero states, which is known as the non-shoot-through state. The diode current is continuous. **Figure 4** illustrates the equivalent circuit. During this time interval, the circuit equations are written as follows:

$$C\frac{dV_{C1}}{dt} = i_B + i_{L2} - i_d \tag{5}$$

$$C\frac{dV_{C2}}{dt} = i_{L1} - i_d \tag{6}$$

$$L\frac{di_{L1}}{dt} = V_{in} - V_{C1} \tag{7}$$

$$L\frac{di_{L2}}{dt} = -V_{C2} \tag{8}$$

Where i_d is the load current going to the inverter.

The average voltages and currents are calculated using the average model between both states. The average voltage in each inductor is equal to zero during one switching cycle (T). From these average inductor voltages equations equated to zero, the capacitor one and the battery voltage relationship are calculated. The average voltage across V_{L1} during each switching cycle is given by

$$(V_{in} - V_{C1})(1 - D) + (V_{in} + V_{batt})D = 0 \tag{9}$$

The average voltage across V_{L2} during each switching cycle is given by

$$(-V_{batt})(1 - D) + V_{C1}D = 0 \tag{10}$$

Eqs. (9) and (10) can be solved simultaneously with respect to V_{batt} and V_{C1} to obtain that

$$V_{pn} = V_{batt} + V_{C1} = \frac{1}{1 - 2D}V_{in} \tag{11}$$

For the active SPWM time, (T1), the effective inverter DC rail voltage can be calculated using the circuit configuration shown in **Figure 4** as follows

$$V_{batt} = \frac{D}{1 - 2D}V_{in} \; and \; V_{C1} = \frac{1 - D}{1 - 2D}V_{in} \tag{12}$$

$$-i_{batt} = i_{L2} - i_{L1} \tag{13}$$

To derive the line to neutral peak voltage equation, the traditional inverter relationship is used, ($\hat{v}_{ac} = \frac{V_{pn}}{2}M$). By combining this equation with Eq. (11), it is easy to calculate the AC output voltage as Eq. (14)

$$\hat{v}_{ac} = \frac{V_{batt} + V_{C1} = \frac{1}{1-2D}V_{in}}{2}MB \tag{14}$$

2.3.1 Circuit analysis for battery parallel with C_1

In this configuration, the battery will be placed across the capacitance C_1 as shown in **Figure 3(b)**. The analysis conducted in the previous section will be repeated for this case to get the relevant equations [10].

2.4 Shoot-through state mode

This mode will make the inverter short circuit via any one phase leg, which is known as the shoot-through state. Consequently, the diode is turned off as the anode to cathode voltage is reversed. The equivalent circuit is shown in **Figure 6**. In this mode, the equations governing the circuit operation are formed as follows:

$$C\frac{dV_{C1}}{dt} = i_B - i_{L2} \tag{15}$$

$$C\frac{dV_{C2}}{dt} = -i_{L1} \tag{16}$$

$$L\frac{di_{L1}}{dt} = V_{in} + V_{C2} \tag{17}$$

$$L\frac{di_{L2}}{dt} = V_{C1} \tag{18}$$

2.5 Non-shoot-through state mode

In this mode, the inverter is free to follow any one of the six active (non-zero) states and two conventional zero states, which is known as the non-shoot-through state. The diode current is continuous. **Figure 7** illustrates the equivalent circuit. During this time interval, the circuit equations are written as follows:

$$C\frac{dV_{C1}}{dt} = i_B + i_{L2} - i_d \tag{19}$$

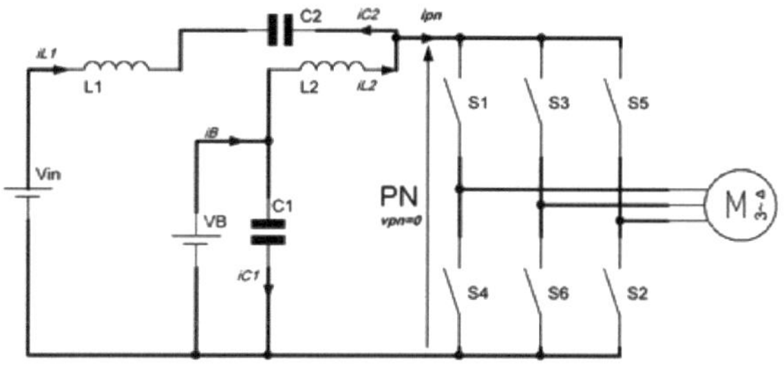

Figure 6.
Shoot-through equivalent circuit [10].

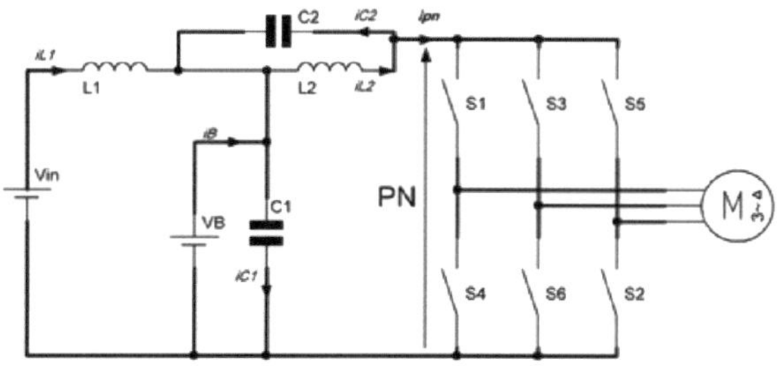

Figure 7.
Non-shoot-through equivalent circuit [10].

$$C\frac{dV_{C2}}{dt} = i_{L1} - i_d \tag{20}$$

$$L\frac{di_{L1}}{dt} = V_{in} - V_{C1} \tag{21}$$

$$L\frac{di_{L2}}{dt} = -V_{C2} \tag{22}$$

Therefore, the average voltages and currents have the relationships as

$$V_{pn} = V_{batt} + V_{C2} = \frac{1}{1-2D}V_{in} \tag{23}$$

$$V_{batt} = \frac{1-D}{1-2D}V_{in} \text{ and } V_{C2} = \frac{D}{1-2D}V_{in} \tag{24}$$

$$i_{batt} = i_{L2} - i_{L1} \tag{25}$$

The voltage V_{C1} of capacitor C_1 will be approximately equal to the battery voltage V_{batt} if the voltage drop on the battery's internal resistance is ignored. Thus, from Eqs. (23) and (24), the DC-link peak voltage V_{pn} will be

$$V_{pn} = 2V_{batt} - V_{in} \tag{26}$$

The output power of the inverter can be controlled by manipulating the desired output voltage, while the output peak phase voltage of the inverter is the same like previous mode.

3. Design and control of battery-assisted QSZI

The power flow of the complete system is affected completely by actual conditions of battery state of charge. The control algorithm is branched into three main categories: battery management system; PV MPPT and motor control. A battery must be protected against overcharging or being depleted. This is achieved by adding a battery management system. PV MPPT is used to guarantee operation at MPPT condition. The motor control will achieve the power balance between the three main systems, namely PV panels, battery and pump.

3.1 Battery management system

Two cases are to be discussed according to the battery SOC:

The first case: the battery is in charging mode. The state of charge is at the possible minimum SOC level. The battery power and load power as the dependent variable are to be determined. It is assumed that proper sizing of the system parameters has been carried out. As a result, the power generated from PV will be less than the combined ratings of the load and battery together. The MPPT tracking system determines the duty ratio. The modulation index is selected based on the battery state of charge.

The second case: the battery is in discharging mode. The state of charge is at the possible maximum SOC level. The battery power and load power as the dependent variable are to be determined. Again, it is assumed that proper sizing of the system parameters has been carried out. As a result, the power generated from PV will be less than the combined ratings of the load and battery together. Similar to the first case, the MPPT tracking system determines the duty ratio, the modulation index is selected based on the battery state of charge.

The detailed algorithm of the battery management is shown in **Figure 8**. This figure is built to show the complete flowchart of the system when battery parallel with capacitor C_1.

3.2 PV MPPT

The most famous method followed in MPPT calculations is perturb and observe. A small perturbation in the system is made. Accordingly, the effect of this perturbation on the power is observed. The main purpose is to maximize the power. If the response is positive (power increases), then apply the same amount of perturbation. The perturbation will be reversed if the response is negative (power decreases). This iteration continues until the system's maximum power point is reached. Shoot-through duty ratio affects the system voltages as explained in Section 2, and so it is used as perturbation element. It is clear from Eq. (23) and Eq. (24) that there is a negative relation between the shoot-through duty ratio and the input voltage. This input voltage is the PV array voltage in the system under study. So, if it is required to increase (decrease) the PV array voltage, the duty ratio has to be decreased (increased).

Good performance in MPPT determination is achieved by the perturbation algorithms. It is simple to implement and can work even if the PV array information is missed. However, an action must be taken to stop the perturbation action when the system reaches the maximum power point. This is because the system continues to perturb around the maximum power point. This may affect the output waveforms quality and decrease the performance of the system generally.

The algorithm is implemented by setting an initial value of 0.05 duty ratio and a step of 0.001, and then, at each measurement the duty is changed by the step 0.001 until the algorithm reaches the final value. Flow chart of the algorithm is illustrated in **Figure 9**.

3.3 Motor: pump set control

In pumping system, the continuity of the water flow, rather than the speed requirement, is considered to be more important. The more power consumed, the more water flow obtained. Power consumed by the load (motor–pump) comes from the power offered by both sources PV and storage batteries.

The motor is loaded by the water pump, and the load torque T_L can be expressed mathematically by

$$T_L = k\omega_r^2 \tag{27}$$

Where k is the pump constant, and ω_r is the rotor speed.

The algorithm is implemented by setting an initial value of 0.8 for the modulation index (M) and a step of 0.001, and then, at each measurement the modulation index is perturbed by the step 0.001. The value of M is linearly proportional to the voltage and

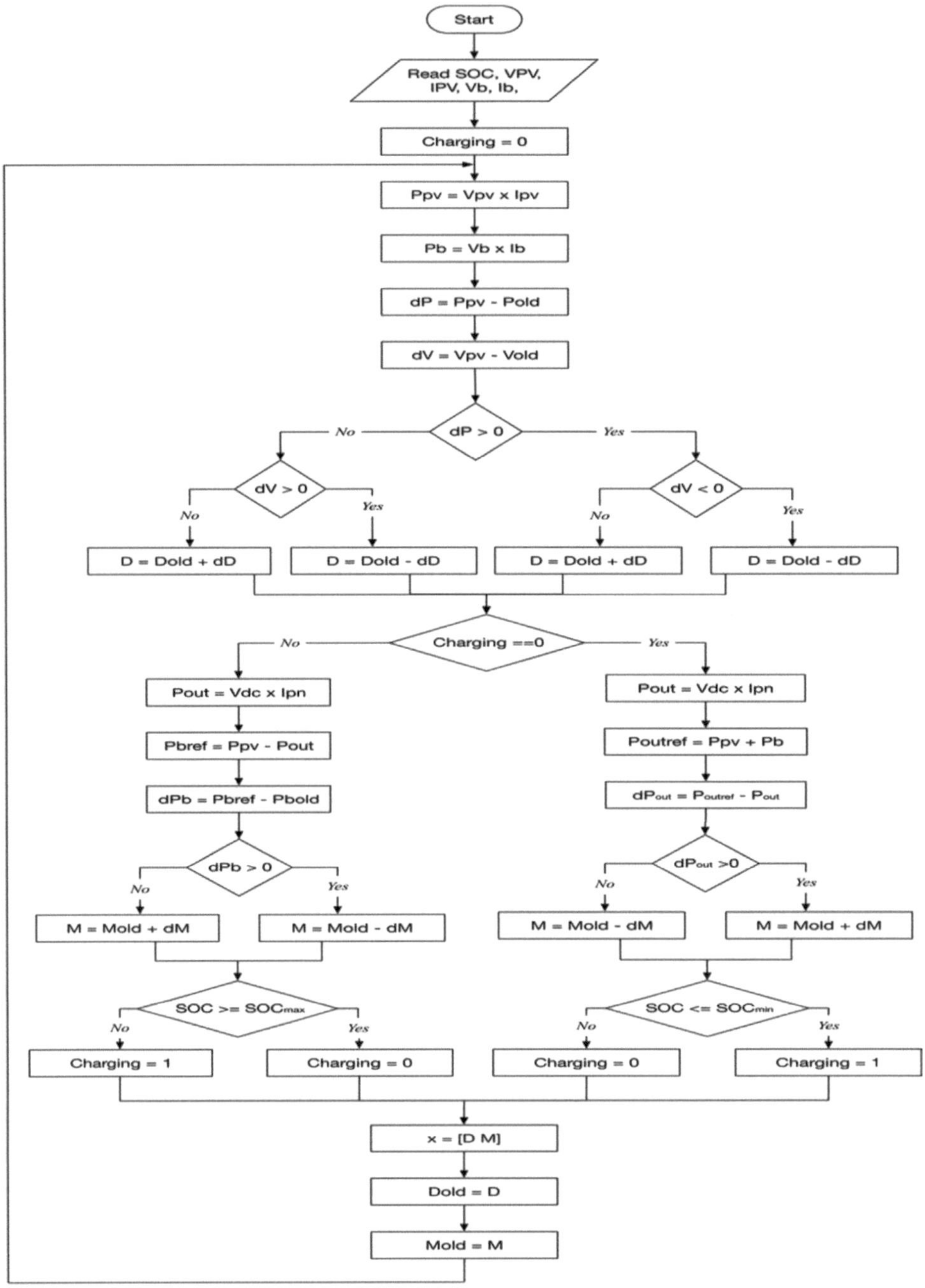

Figure 8.
Complete flowchart of the system when battery parallel with capacitor C_1.

frequency of the motor. Hence, rated frequency (60 Hz) is proportional to M = 0.8, and if the modulation index decreased, the frequency is linearly decreased. As a consequence, the voltage is also decreased (460 V is proportional to 60 Hz) until the algorithm reaches the final value.

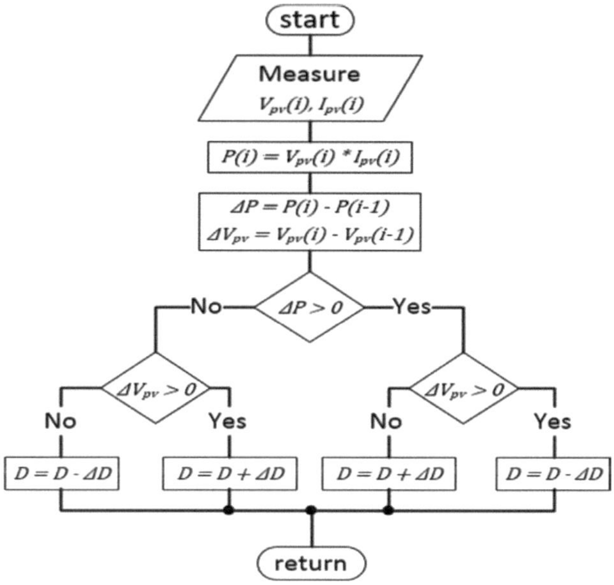

Figure 9.
Flowchart of P&O algorithm using duty ratio as perturbation [11].

In scalar control, the control of the magnitudes of voltage and frequency leads to the control the torque and the magnetic flux of the motor. Induction motor has an inherent coupling effect. Motor-induced torque and magnetic flux are functions of voltage and frequency. Varying the applied voltage will affect both the induced torque and the magnetic flux. The dynamic performance of the scalar control got a weak point. However, scalar control implementation is simple. Then by using V/f control speed can be directly controlled. By controlling the torque and speed of the pump, the power consumed by the system is controlled [12].

The power consumed by the induction motor is given by

$$P_{in} = 3V_sI_s coscos\ \varphi = k\omega_r^3 \tag{28}$$

where V_s is the stator phase voltage,
I_s is the stator current and
$cos\ \varphi$ is the power factor

4. Simulation results

In this section, a model representing the PV-pumping system will be simulated to verify the control technique. The model was developed using MATLAB/Simulink environment as shown in **Figure 10** [12].

System parameters values at rated power conditions are listed in **Table 1**. Induction motor values have been taken from the MATLAB/Simulink. The supply frequency is 60 Hz. The PV array is selected to suit the motor loading condition given

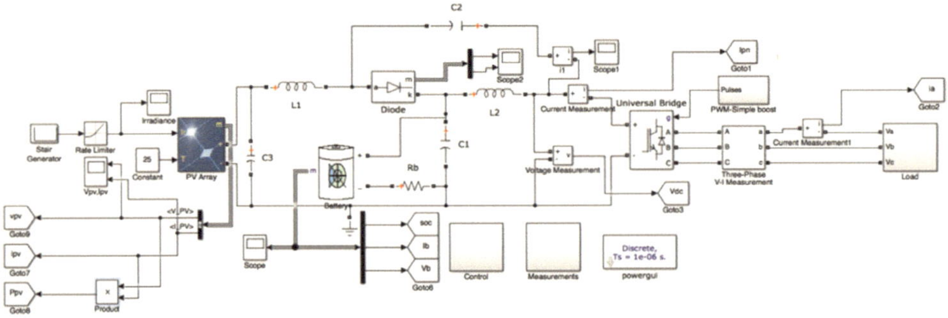

Figure 10.
Simulink model of the PV-pumping system.

Parameter			Value
PV array	@ 1000 W/m^2	V_{mpp}	611 V
		I_{mpp}	7.5 A
		P_{max}	4500 W
		V_{OC}	830 V
		I_{SC}	8.25 A
	@ 500 W/m^2	V_{mpp}	630 V
		I_{mpp}	4.2 A
		P_{max}	2400 W
		V_{OC}	780 V
		I_{SC}	4.25 A
QZSI		L	10 mH
		C	400 μF
Induction motor		Rated RMS V_{L-L}	460 V
		Rated frequency	60 Hz
		Rated power	4000w

Table 1.
System parameters' values at rated power.

that the PV array is working at its maximum power. In the simulation, the PV array is treated as one unit. In reality, number of panels, power rating of each panel, and how they are electrically connected should be determined. The irradiance is assumed to be changed from 1000 to 500w/m^2 after 5 seconds. The simulation sample time is selected to be 1 μs, given that the switching frequency is 10 KHz, a shoot-through duty ratio resolution is 0.01. Each simulation run takes 2–3 hours. This computation time could be reduced if a faster PC is available.

4.1 MPPT performance investigation

Figure 11 indicates the calculated PV performance parameters (voltage, current and power) with step change of duty ratio equals 1e-6. It is seen from the

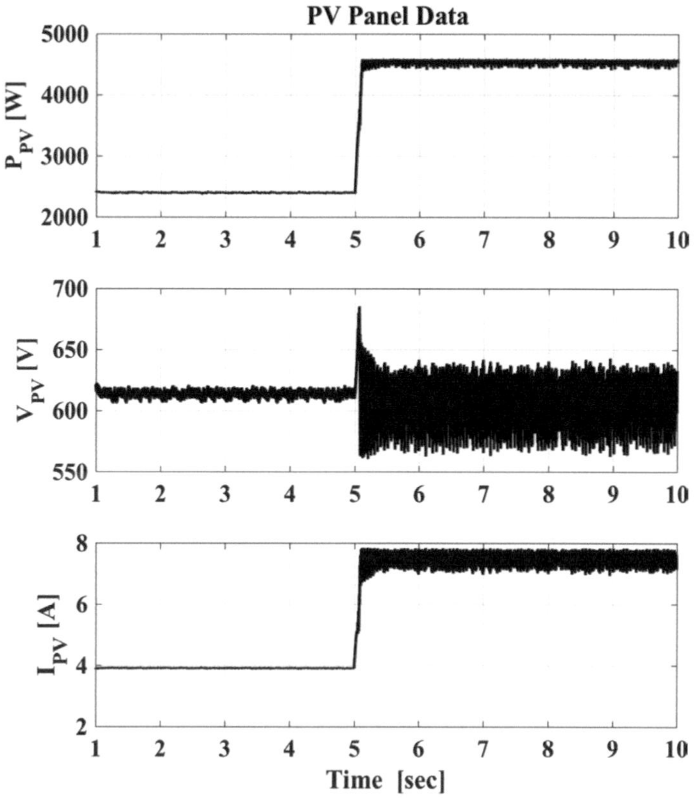

Figure 11.
PV power, voltage and current with P&O algorithm.

figure that the system can reach MPPT, even, at different insolation as recorded in the power curves. A good system components design leads to the fact that the rated PV panel power is larger than the rated power of the motor–pump system and battery charger power. The system shows that the solar panel is capable, at MPPT, to generate 4.5 kW at 1000 W/m^2 and 2.4 kW at 500 W/m^2. The perturb-and-observe controller suffers from a minor problem. The system tends to oscillate around the maximum power point, which is observed clearly in the power vs. time curve.

4.2 QZSI performance investigation

Figures 12 and **13** show QZSI modulation index variations and the change in the shoot-through duty ratio of the system. Rate of change in shoot-through duty ratio is increased rapidly by the step of change at the fifth second. The control technique keeps always the modulation index to be less than or equal to 1-D. This means that there is no overlapping between both controls. The increase of the irradiance of the system leads to increase in both the modulation index and the shoot-through duty ratio.

Voltages across capacitors C_1 and C_2 under different irradiance conditions are shown in **Figure 14**. **Figure 15** records the inductor currents in L_1 and L_2. Voltage across capacitor C_1 is clamped to a level of around 838 V. This value

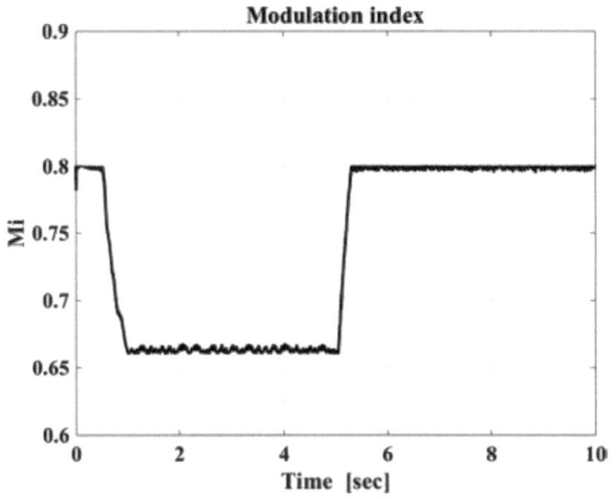

Figure 12.
Modulation index of the system.

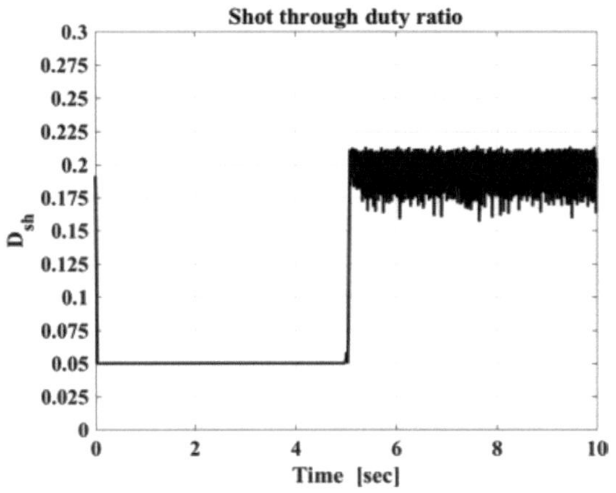

Figure 13.
QZSI shoot-through duty ratio.

includes voltage drop across the resistance of the battery (the battery voltage equals 775 V). Voltage across the capacitor C2 decreases by the decrease of the power. This can be explained depending on the fact that the output voltage is related to the motor power. It is seen that the inductor currents have always positive sign. Due to the existence of the battery, their patterns are not identical.

4.3 Motor performance

Figure 16 shows motor line-line voltage and phase currents. **Figure 17** shows the speed and power of the motor change with respect to changed irradiance; when the power and voltage increase, the speed correspondingly increases which fulfills the constant V/f control technique.

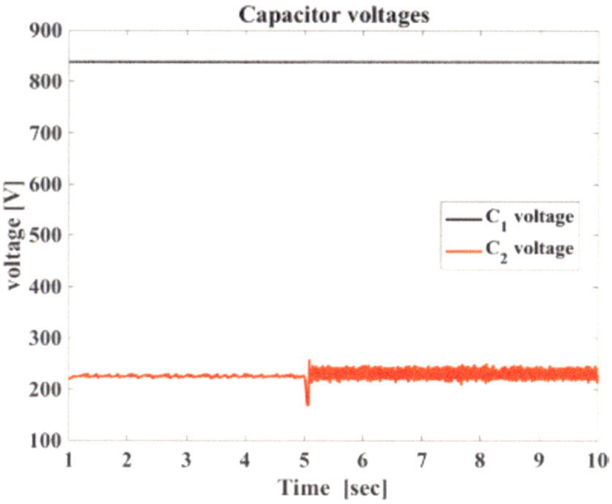

Figure 14.
Capacitor C_1 and C_2 voltages.

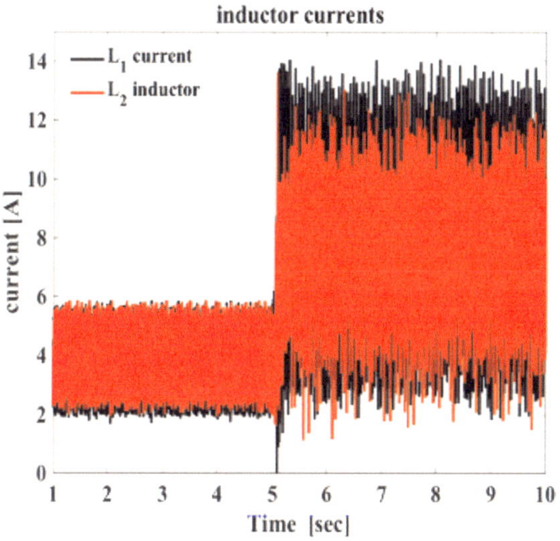

Figure 15.
L_1 and L_2 inductor currents.

4.4 Battery performance investigation

Figure 18 shows the battery performance parameters (state of charge, voltage and battery current). In the discharging mode, the battery state of charge decreases. The battery enters the charging mode and starts charging at the fifth second (when the irradiance increases). The SOC variation, between the two modes, is small. This is due to the short simulation period. The absence of the inductance in the battery current path causes the oscillation in the battery current. This is of course the cost of simple battery charging circuit design.

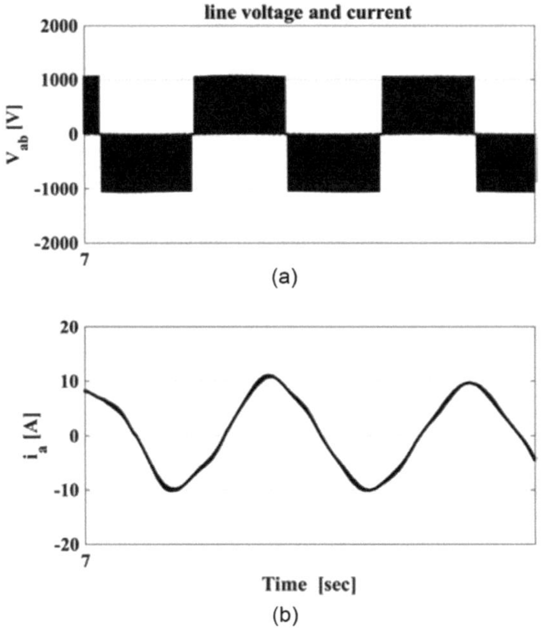

Figure 16.
(a) Motor voltage and (b) motor current.

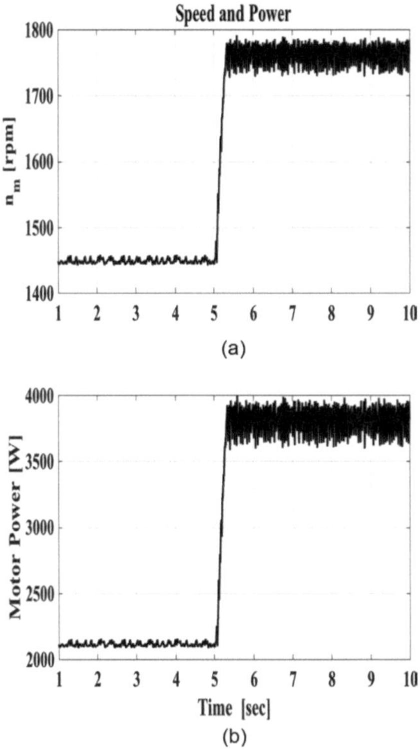

Figure 17.
(a) Motor speed and (b) motor power.

DOI: http://dx.doi.org/10.5772/intechopen.110890

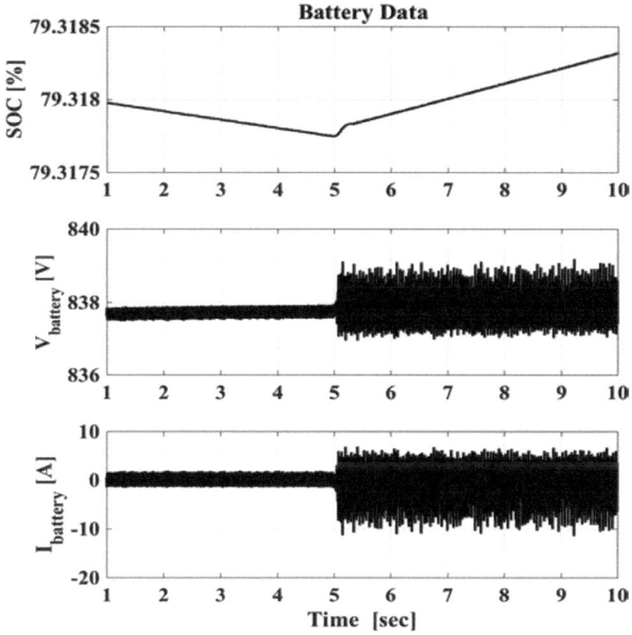

Figure 18.
Battery state of charge, current and voltage.

5. Conclusions

In this chapter, a PV-pumping system was developed. The design is based on battery-assisted quasi Z-source inverter. Traditionally, two-stage converter (boost converter followed by voltage source inverter) is employed in such applications. This work introduces a QZSI mathematical model. The proposed topology added a storage unit (battery). The position of the battery is selected to be shunted either with capacitor C_1 or with capacitor C_2.

In order to guarantee a good PV-pumping system performance, an overall control of the PV-pumping system with QZSI, assisted with a battery, is to be introduced. This algorithm of this controller should be able to force the PV panel to reach MPPT, even, under different conditions. Also, the battery must be allowed to enter the charging and discharging states in the proper timing and according to the system requirements. The inverter is controlled through the choice of the value of the modulation index as well as the shoot-through duty ratio. In order to prevent overlap between both control loops, the modulation index is designed to be less than or equal to (1-D) maximum allowable value. Consistency of simulation results with the analysis is an indication to the accurate system design steps. Simulation results, also, verify the proposed QZSI control technique. The system, under this control algorithm, provides continuous current in the inductors. This charging–discharging profile for the batteries satisfies a longer life time. The disadvantage is that the battery current is rich in ripples. This is cost of the simplicity design of the charging circuit [12].

Acknowledgements

The author appreciates the efforts introduced by Eng. Saad A. Altarfawi.

Author details

Ahmed Alaa Mahfouz
Electrical Engineering Department, College of Engineering, Qassim University, Buraydah, Kingdom of Saudi Arabia

*Address all correspondence to: alaa@qec.edu.sa

References

[1] Peng FZ. Z-source inverter. IEEE Transactions on Industry Applications. 2003;**39**(2):504-510. DOI: 10.1109/TIA.2003.808920

[2] Shen M, Joseph A, Wang J, Peng FZ, Adams DJ. Comparison of traditional inverters and Z-source inverter for fuel cell vehicles - 2007. IEEE Transactions on Power Electronics. 2007;**22**(4): 1453-1463. DOI: 10.1109/PET.2004.1393815

[3] Anderson J, Peng FZ. Four quasi-Z Source inverters. In: 39th IEEE Power Electronics Specialists Conference, Rhodes, Greece. IEEE; 15-19 June 2008. pp. 2743-2749. DOI: 10.1109/PESC.2008.4592360

[4] Cintron-Rivera JG, Li Y, Jiang S, Peng FZ. Quasi-Z-source inverter with energy storage for photovoltaic power generation systems. In: 26th Annual IEEE Applied Power Electronics Conference and Exposition (APEC), Fort Worth, Texas, USA. Vol. 1. IEEE; 6-11 March 2011. pp. 401-406. DOI: 10.1109/APEC.2011.5744628

[5] Sun D, Ge B, Bi D, Peng FZ. Analysis and control of quasi-Z source inverter with battery for grid-connected PV system. International Journal of Electrical Power & Energy Systems. Elsevier. 2013;**46**:234-240. DOI: 10.1016/j.ijepes.2012.10.008

[6] Holland K, Shen M, Peng F. Z-source inverter control for traction drive of fuel cell - battery hybrid vehicles. In: Industry Applications Conference, 2005. Fortieth IAS Annual Meeting. Conference Record of the 2005. Vol. 3. 2-6 2005. pp. 1651-1656. DOI: 10.1109/IAS.2005.1518668

[7] Shen M, Peng F. Control of the z-source inverter for fuel cell battery hybrid vehicles to eliminate undesirable operation modes. In: Industry Applications Conference, 2006. 41st IAS Annual Meeting. Conference Record of the 2006 IEEE. Vol. 4. 8-12 2006. pp. 1667-1673. DOI: 10.1109/IAS.2006.256760

[8] Peng FZ, Shen M, Holland K. Application of Z-source inverter for traction drive of fuel cell–battery hybrid electric vehicles. IEEE Transactions on Power Electronics. 2007;**22**(3):1054-1061. DOI: 10.1109/TPEL.2007.897123

[9] Li F, Ge B, Sun D, Bi D, Peng FZ, Abu-Rub H. Quasi-Z source inverter with battery based PV power generation system. In: Proc. 2011 Int. Conf. Electrical Machines and Systems (ICEMS), Beijing, China, 20-23 August 2011. Vol. 2. pp. 1459-1463. DOI: 10.1109/ICEMS.2011.6073594

[10] Ge B, Abu-Rub H, Peng FZ, Lei Q, de Almeida A, Ferreira F, et al. An energy stored quasi-Z-source inverter for application to photovoltaic power system. IEEE Transactions on Industrial Electronics. 2013;**60**(10):4468-4481. DOI: 10.1109/TIE.2012.2217711

[11] Nedumgatt JJ, Jayakrishnan KB, Umashankar S, Vijayakumar D, Kothari DP. Perturb and observe MPPT algorithm for solar PV systems-modeling and simulation. In: 2011 Annual IEEE India Conference, Hyderabad. 2011. pp. 1-6. DOI: 10.1109/INDCON.2011.6139513

[12] Altarfawi SA, Mahfouz AA. Simple and effective algorithm for battery assisted quasi Z-source inverter for standalone pv-pumping systems, International Journal of Innovative Technology and Exploring Engineering (IJITEE). May 2020;**9**(7):1309-1314. ISSN: 2278-3075 DOI: 10.35940/ijitee.G4877.059720

Chapter 4

Energy Storage System (ESS) in Residential Applications

Ya LV

Abstract

This chapter looks into application of ESS in residential market. Balancing the energy supply and demand becomes more challenging due to the instability of supply chain and energy infrastructures. But opportunities always come with challenges. Apart from traditional energy, solar energy can be the second residential energy. But solar energy by nature is intermittent and available under solar irradiance only, so we need a solution to harvest all the solar energy generated in daytime and use it for supplying household load at high demand or backup at power outage, this solution is ESS. Most residential ESS solutions are Lithium-ion battery (LiB) based due to its high energy density and small footprint. But degradation of LiB system is quite sensitive to application conditions like temperature, and the lifespan of most LiB systems is between 10 and 20 years. Looking at future trend, the residential renewable energy solution (RRES) will become more flexible, compatible and reliable. Digitalization, as development trend, will enable end-users remotely monitor and manage different operation modes of RRES. The RRES suppliers will also offer most economic operation plan for end-users given their geography locations and household energy consumption habit.

Keywords: ESS, RRES, solar energy, LiB storage system, digitalization

1. Introduction

Energy supply shortage and environment deterioration are two serious issues that deserves our attention. Our daily life continuously consumes electrical energy, the main electricity supply in US come from natural gas, nuclear, and coal in 2020 according to US Energy Information Administration. These supply sources all pollute environment by generating greenhouse gas and poisoning plants and animals in nature. This is driving us to explore clean energy, such as solar energy, wind energy and hydropower. The renewables are the fastest growing sector and comprise the biggest portion of new energy sources deployed on the grid.

By generating electricity from converting hydropower, wind energy and solar energy, the development of renewable energy originates from 19th century [1]. Going through two-decades efforts, in different countries inclusive of China, US and UK, large Dams were built up for harnessing hydropower, different scales of wind farms and solar plant were developed for converting wind energy and solar energy into electricity.

But in residential applications, currently solar energy is the sole renewables for being deployed. In 2021, deployment of photovoltaic (PV) modules was expanding rapidly in various countries, amounted 433 MWp in Singapore [2], 54.9 GWp in China [3], 730 MWp in UK [4], 25.9 GWp in EU [5], 23.6 GWp in US [6], 12 GWp in India [7]. Until 2022, PV module design efficiency stands at 15–27%, in which mono-crystalline module efficiency can reach 22–27%, polycrystalline and thin film module efficiency is between 15% and 22% [8]. Researchers worldwide are dedicated to push boundaries of efficiency in solar cell and PV module by studying materials and processes. In addition, PV modules' application performance is inevitably affected by weather conditions like solar irradiance and shading. As a result, PV modules always generate electricity in an intermittent way.

ESS is the effective solution of storing intermittent electricity generated by PV modules. In residential applications, the power flow within household is within 7.36 kW for single-phase, so the residential ESS power is in similar scale, in which Li-ion battery (LiB) based ESS is the most popular solution.

In this chapter, we will begin with introduction of residential renewable energy solution (RRES), next focus on understanding of LiB based ESS, and end with discussion of future solutions and technical trends in RRES and residential ESS.

2. What is residential renewable energy solution (RRES)?

The RRES is an energy optimization solution for household users that integrates solar energy, power grid and ESS altogether. Without the RRES, we pay higher electricity bills at grid peak hours and lose power supply at grid outage times. The invention of RRES cuts our electricity bills by selling back solar energy and guarantees a backup power supply at blackout times. The residential ESS functions to store intermittent electrical energy from PV modules and provide power supply for backup loadings. The current RRES market is dominated by players from APAC region, North America and Europe, such as LG Electronics, Tesla, Huawei, Enphase, and Siemens [9].

In the RRES, PV modules and battery storage can be coupled in a DC or AC way, as shown in **Figure 1**. The solar energy will be transmitted to grid in default mode, the timing of when to charge the battery storage is determined by the optimized e-bill. If DC coupled system, PV modules charge the battery storage via a maximum power point tracking (MPPT) solar charge controller and DC-DC converter that is embedded into the Hybrid bi-directional Inverter; in an AC coupled system, battery storage will be charged via the DC-AC Inverter.

Let's look at an example of deploying a RRES system in Singapore. The daily electricity system demand can refer to **Figure 2** [10], it can be seen that the usual grid peak hours are 18:00–24:00. The daily solar irradiance curve in Singapore can refer to **Figure 3**, the peak solar beam irradiance occurs between 11:00 and 16:00. By comparing **Figure 2** with **Figure 3**, we find that there are little or no solar energy at the grid peak hours, therefore, we need battery storage to store the solar energy generated from daytime. At grid peak hours, the battery storage, via an inverter, can provide energy supply to household loadings or feed energy back to grid for reducing the electricity bill. Moreover, the battery storage together with the inverter can work as uninterruptible power supply (UPS) to supply power at blackout or brownout times, this will become an emergency measure and also prevent equipment damage or data loss.

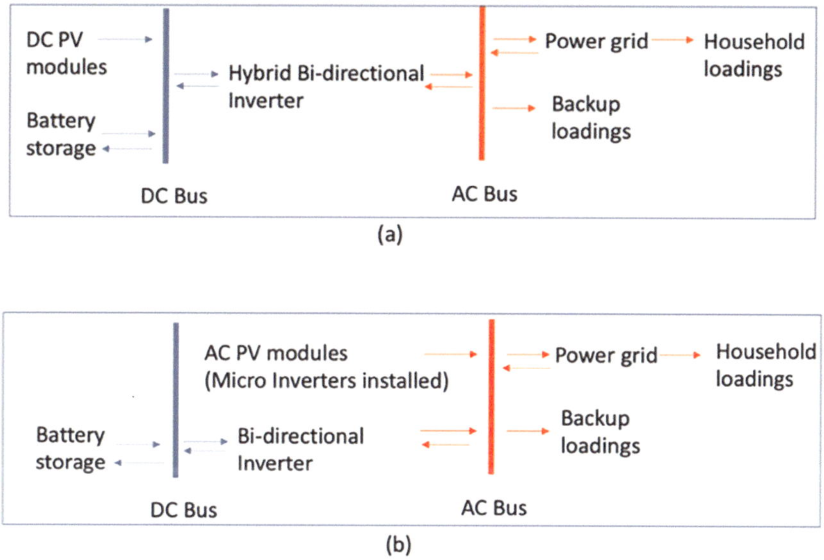

Figure 1.
The RRES solution diagram (a) DC coupled solution and (b) AC coupled solution.

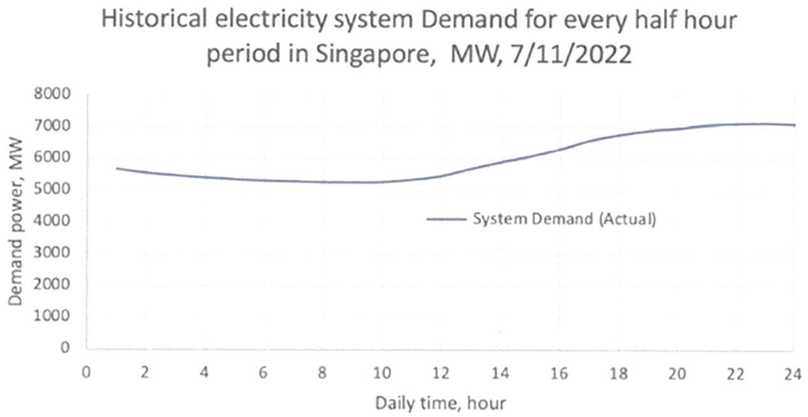

Figure 2.
Singapore electricity system demand on 7/11/2022.

3. About residential ESS-LiB storage

In **Figure 4** we compare the development history of LiB [11] to that of power electronics. The research of LiB began 10 years later after the Bell lab's development of Silicon transistor, similar to the development of different types of transistors, the LiB experienced different recipes of the cathode materials. Nowadays, the solid-state battery (SSB) based on solid electrolyte has entered the residential market. Furthermore, the digital control and intelligent drive of power electronics originated in 1990s can represent one of the technology trends in future LiB system as well.

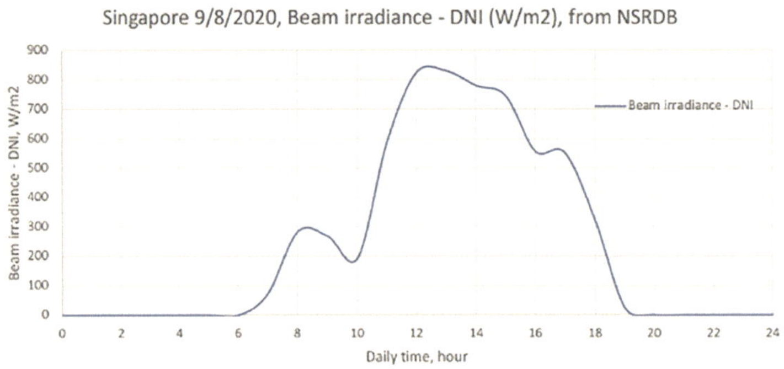

Figure 3.
Solar beam irradiance in Singapore on 9/8/2020, data from NSRDB.

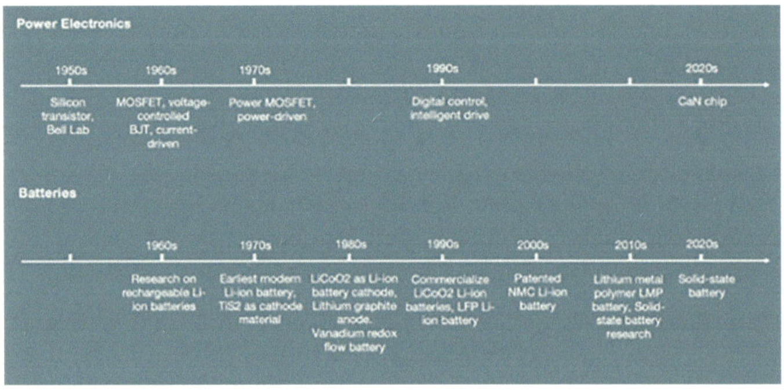

Figure 4.
Main milestones in development of power electronics and that of LiB.

The LiB is composed of a cathode, anode, electrolyte, separator and two current collectors. As indicated in **Figure 5**, the cathode material provides positively charged lithium ions and gets connected with the positive current collector; the anode material stores lithium ions and gets connected with the negative current collector; the electrolyte stays in between the cathode and anode for transporting lithium ions; the separator blocks the transportation of electrons between the two electrodes. During the charging process of LiB, lithium ions are released from the cathode and transported to the anode to be stored. With an external loading, the LiB can be discharged with the flow of electrons in an external circuit via the transportation of the released lithium ions from the anode to the cathode.

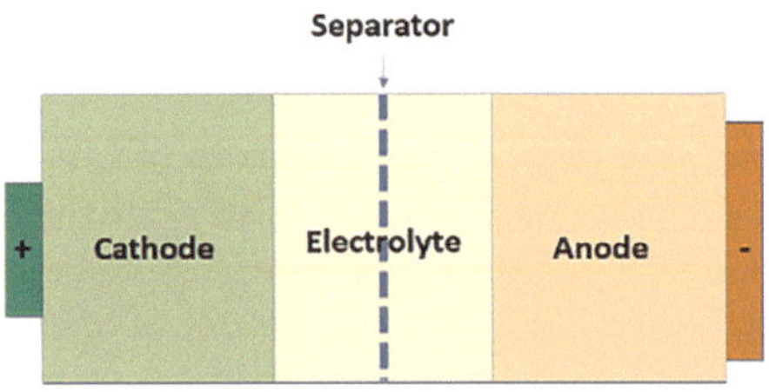

Figure 5.
Simplified diagram for LiB composition.

3.1 LiB storage system integration process

LiB storage system integrates different quantities of LiB cells in a functional and safe way. There are seven design perspectives to understand this integration process, that can be abbreviated as 'TEAMFCI.'

'T' is a thermal management solution. Because the round-trip efficiency of LiB system is around 92-95% in market products, there will be some amount of heat to generate in LiB system, but the LiB temperature is a critical stress factor that affects its working efficiency and reliability which affect lifespan, so we need to tackle the generated heat with efficient thermal management methods. In a LiB storage system shown in **Figure 6**, there are usually passive and active thermal management methods. Passive methods rely on heat conduction and natural heat convection whereby the heat is conducted from the LiB cells to system enclosure surfaces, then the heat is exchanged in a natural way between the system enclosure surfaces and the environment. Active methods include forced heat exchange between the LiB cells and a flowing fluid whereby the heat will be further exchanged out to ambient from the flowing fluid using fans/heat exchangers. The heat transfer efficiency between different objects can be described with thermal impedance which consists of both thermal resistance and heat capacity. At steady state, during long duration heat transfer, the heat capacity can be ignored in the heat transfer path. However, at transient states of short-duration heat transfer, the heat capacity needs to be considered. In the residential LiB storage system, the heat capacity of LiB cells cannot be ignored due to the cells' dominated mass in the system. The thermal network of passive and active thermal management solutions in such system can be simplified in **Figures 7** and **8**.

'E' is electrical connectivity. A configuration design of LiB storage system generates certain system-level voltage and current by connecting different number of cells in parallel and in series. Usually, the cells in parallel will be connected via bus bars, followed by serial connectivity via bus bars/electrical cables. In addition, there will be connectivity of data acquisition cables and power supply cables between cells and the battery management system (BMS), as well as connectivity of power cables between cells and the system terminals. BMS is a power electronic device that provides four main function blocks: (1) data acquisition for cell-level voltage, current and temperature; (2) cell-level voltage balancing during charging process to maximize the usable capacity in system; (3) functionality safety control for cell-level protection of overvoltage, undervoltage, overcurrent, overtemperature and under temperature; (4) communication with master control devices like energy management system (EMS)/master control computer/master BMS via protocols like CAN bus/Modbus.

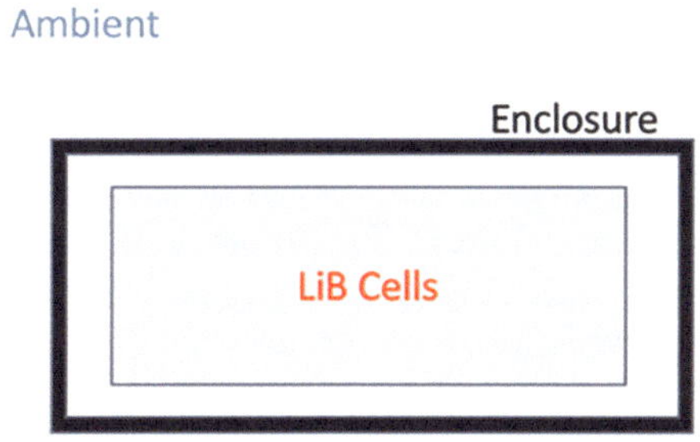

Figure 6.
Simplified LiB storage system.

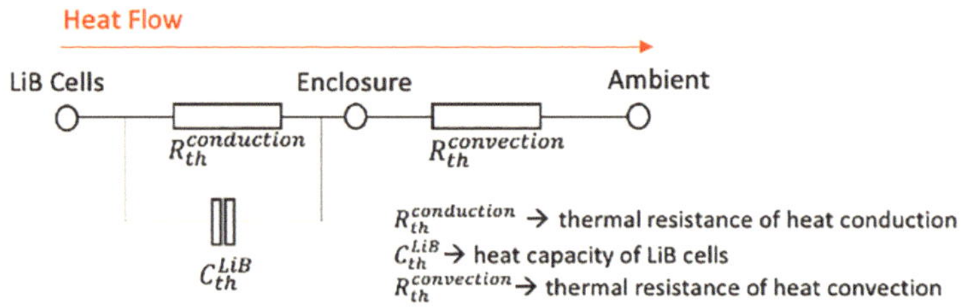

Figure 7.
Foster thermal network model for passive thermal management solution.

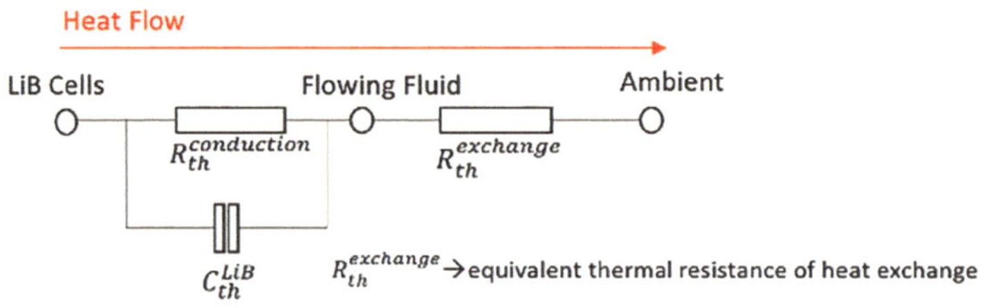

Figure 8.
Foster thermal network model for active thermal management solution.

'A' means algorithm of state of charge (SoC) and state of health (SoH) calculation in the BMS. With this algorithm, the BMS can record cell-level real-time charging/discharging status and calculate system health status along the lifespan for understanding the throughput energy. This algorithm can be further developed for serving preventive maintenance and predictive maintenance, in order to prolong the workable lifespan of LiB system via increasing the throughput energy.

'M' indicates mechanical design. In LiB storage systems, mechanical design mainly covers design of internal frame that is used for holding battery cells steadily and an external enclosure that demands good sealing and corrosion resistance.

'F' represents fire-safety design. All the plastic materials chosen for LiB storage system are required to be UL94 V-0 rated, so the burning can stop within 10 seconds during fire propagation.

'C' means certification. For LiB storage systems, there are at least three crucial certifications: (1) underwriters' laboratories (UL) 1642, the standard for Lithium battery-specific testing which tests the risk of fires and explosion. The equivalent international electrotechnical commission (IEC) standard is IEC 62133; (2) UL 1973, the standard for batteries for use in stationary/vehicle auxiliary power/light electric rail applications which certifies the capability of battery systems under normal and abnormal conditions. The equivalent IEC standard is IEC62619; (3) UL 9540, the standard for ESS to meet industry and regulatory needs. The equivalent IEC standard is IEC 62933-5. UL is an accredited standards developer in the US and Canada [12]. IEC works with accredited laboratories in Europe, Middle East and Asia [13].

'I' indicates an industrial design that places an impression to end-users on the whole system, this part becomes necessary for products that are targeted at residential applications.

3.2 LiB storage system performance indicators

The LiB storage system performance can be evaluated by five indicators, abbreviated as '4CS.'

The first 'C'—C-rate for continuous and maximum operation. The C-rate describes the charging/discharging speed, 1C-rate means fully charging/discharging the system within one hour, 2C-rate indicates double speed and half time of 1C-rate.

The second 'C'—capacity. A residential LiB storage system, as an energy supply device, seeks to have a small footprint and be able to wall-mounted. Moreover, low noise level is preferred in such LiB storage system, so the thermal management solution is prone to take passive measures, such as inside heat conduction and outside natural convection, even taking active measures, there won't be strong-forced fluid flow. As a result, in residential LiB storage systems, the continuous charging/discharging C-rate is usually constrained up to 0.5C, to maintain its workable lifespan and minimize thermal runaway risk. In residential applications, the circuit breakers of houses are usually 16A rated and 32A rated, based on the single-phase voltage of 230 V, the drained power from the grid is limited to 3.68 and 7.36 kW respectively, correspondingly, the residential LiB storage capacity per house is around 10 kWh. The 10 kWh can be one LiB storage module or a combined system of several modules, this configuration depends on the module's integration design in terms of 'TEAMFCI.'

The third 'C'—cyclability. Lifespan of a LiB storage system is measured by its operation cycles. Manufacture of LiB cells can provide testing report of cell-level operation cycles under certain conditions, for instance, 6000 cycles under room temperature and 0.5C-rate charging and discharging. Integrating LiB cells to a LiB storage system, generates a cell-to-system cyclability loss caused mainly by the SoC difference and temperature non-uniformity among all cells. Moreover, the LiB storage system will fade along with its lifespan via cycling degradation and calendar degradation. The cycling degradation can be measured by system SoH that describes the change of system throughput energy by comparing the current value to the beginning-of-life (BoL) value. The key stress factors associated with system SoH are temperature, SoC and current. Their respective influence is elaborated as below:

1. Temperature—The optimal operating temperature of LiB cells is between 15 and 35°C, in order to guarantee its working reliability and maintain its maximum cyclability [14]. Temperature effects exist in both calendar aging and cycling aging modes. At high temperature [15], the solid-electrolyte-interfacial (SEI) layer at the anode will deteriorate and gradually dissolve into the electrolyte The damaged SEI layer will be restored from the side reactions between the exposed active anode material and electrolyte, but this process will induce the difficult intercalation and lower ionic conductivity at the anode. The same degradation mechanism occurs at the cathode side with solid-permeable-interface (SPI) layer This can cause structural damage to the active cathode material. At low temperature, there is sluggish electrochemical reaction [16], which will induce output power degradation and irreversible capacity loss. The relationship between the capacity fading and temperature can be represented in Arrhenius equation [15], $A = A_0 * \exp\left(-\frac{E_a}{RT}\right)$, Where A is the amount of capacity fading, A_0 is the pre-exponential term, E_a is the activation energy, R is the gas constant and T is the temperature in Kelvin. This equation is applicable to describe both calendar fade and cycling fade. The activation energy E_a decreases at higher battery SoC, which means that capacity fades faster at higher SoC.

2. SoC—LiB SoC represents the available capacity in the LiB cells. At high SOC [14], more electro-chemical reactions will take place and the SEI layer will grow faster, as well as the aggravating self-discharge. In cycling mode, the SoC effect on LiB cells degradation cannot be described well owing to short maintaining time at different SoC levels. Usually, there is an advised operation SoC range in LiB storage system, namely the depth of discharge (DoD), in order to minimize the negative impact on system SoH.

3. Current—Current inevitably affects system SoH given the generated ohmic heat power and thus makes the current effect become part of the temperature impact.

Based on above analysis of key stress factors, the LiB storage system SoH can be calculated and predicted if the system temperature can be obtained and predicted as well as an equivalent system-level activation energy. A semi-analytic methodology (SAM) [17] can be utilized to quickly calculate LiB storage system temperature. This SAM is extracted by studying active thermal management in the LiB storage system. In the active thermal management, the main heat exchange happens at the interface between solid and fluid. This SAM article [17] analyzed thermal networks for both steady-state and transient-state heat transfer situations, as well as studying the influence of different input parameters on system temperature. Compared to complicated computation resources and long computation time in numerical simulation, this SAM can calculate the thermal outcome based on variable inputs within a few minutes. This SAM can be part of an algorithm for SoH calculation and prediction in LiB storage systems while also potentially optimizing the system operation cyclability and workable lifespan.

The fourth 'C'—cost. The cost breakdown in LiB storage system are mainly from battery cells, BMS, and system integration. On top of concern of battery cost and BMS cost, the cost of system integration needs to be balanced and considered, especially the cost of thermal/electrical/mechanical solutions. In additions, the cost of system lifespan and that of system scalability are required to be counted into the calculation. In residential markets, some low-price LiB storage systems have not covered scope of long lifespan and system scalability.

The 'S'—safety risk. Temperature of LiB storage system is crucial given its direct triggering of thermal runaway. Thermal runaway of LiB cells can be triggered at about 85–120°C owing to the solid-electrolyte interface decomposition, followed by battery separator melting at about 130°C, then the cathode decomposition and electrolyte oxidation will occur at above 150°C, this is the timing of entering thermal runaway and catching fire. Thermal runaway in LiB storage system can be caused by factors like internal short circuit/functionality safety malfunction/overheating/mechanical damage of batteries. The Lithium Ferro Phosphate (LFP) cathode-based LiB is more electrochemical stable and more suitable for residential applications compared to the Nickel Manganese Cobalt Oxide (NMC) cathode-based LiB. The fire-safety design in LiB storage system covers fire propagation mitigation. Meanwhile, the safety evaluation of LiB storage systems can be certified with UL 9540/IEC 62933-5 whose testing scope covers system thermal runaway, so as to better prevent from danger.

4. Future of residential ESS

The whole RRES is developing towards higher and higher flexibility, compatibility and scalability.

1. The high flexibility means that a friendly user interface can be accessed to monitor and manage system operation modes, in order to optimize electricity bills to the biggest extent. This can include choosing solar irradiance peak hours to charge LiB storage and selecting grid peak hours to discharge LiB storage. By collecting and analyzing the big data from end-users, different user portrait models can be generated for realizing automatic optimization of system operation modes.

2. The high compatibility aims at including more user demand into this eco system, like heat pump and EV chargers. Apart from energy management system (EMS), there will be a central/distributed Gateway solution that is embedded with data analytics models. This Gateway solution can communicate among different products in the eco system.

3. The high scalability means to build up a bridge between the RRES and utility renewable energy system (URES), the RRES and URES can become each other's power source or power demand, which helps us to connect the distributed RRES dots and knit a bigger eco network.

Based above understanding of future RRES, we can sense that the future residential ESS will become more and more intelligent and diverse.

4.1 Future of LiB storage system

In LiB storage solution, there are two key constituents, one is LiB cells, the other is BMS. The future of LiB cells and that of the BMS represents the future of LiB storage solution.

Climate change affects all aspects of civilization and where LiB are concerned the LiB cells that can sustain at higher average temperature will be one of the future needs. The other trend of future LiB storage in residential markets is to repurpose the e-mobility LiB storage systems. The LiB share in e-mobility market will reach up to 2333 GWh by 2030 [18], Worldwide, it is expected to produce over 2 million metric tons of used batteries per year by 2030 [19]. The retired LiB storage systems still retain 70–80% of BoL throughput energy. Compared to directly recycling the retired systems, repurposing them into less-challenging applications will be better way in terms of environmental friendliness and supply chain stability. This repurposing strategy will facilitate the spread of RRES over the world as well. There are a plethora of research work on the repurposing methodologies. The sequence starts from quality checking and goes through classification, integration and commissioning. The quality checking includes both cosmetic checking and functionality validation. Because the SoH of retired system is usually limited by the weakest cell, the identification of healthy cells in a system can help to improve the throughput energy via replacing the weak cells or bypassing them in the repurposed electrical circuit. A non-invasive battery grading algorithm can be found in reference [20]. In this proposed battery grading algorithm, cell-level historical voltage data as a reflection of cell degradation is investigated as well as cell-level historical maximum temperature data calculated with the developed SAM methodology [17]. The LiB cells in the retired system can be graded to three levels by analyzing their historical voltage fluctuation and working temperature variation. The rationality of this algorithm was validated in the 'Battery Grading Algorithm' work by Ya et al. [20].

An intelligent BMS that can sync real-time cell-level data to remote cloud and accept remote operation command will be the future trend. The boundary between BMS and EMS will be more and more blurred over time. The existing BMS solutions include four function blocks. The first function block is to collect cell-level sensing data from sensors of voltage, current and temperature. The second function block is to balance the voltage among different LiB cells in order to optimize lifespan of the whole system, otherwise the lifespan will be affected by the weakest cell which experiences over charge or over discharge. The third function block is to safeguard the whole system by monitoring the collected data form the first function block. Usually there are two-levels warnings if the threshold values of voltage/current/temperature are triggered, if the warnings are maintained, the BMS will force the system to power off via relays. The fourth function block is for the BMS to communicate with external system like EMS or the other BMS. In future BMS solutions, it is hoped to sync real-time cell-level data to remote cloud for analytic models, and the weak battery cells can be bypassed via accepting remote command once it is diagnosed as unhealthy or abnormal, in this way, lifespan of the whole LiB storage system will be extended to the biggest extent.

4.2 Future of other residential ESS

Following the diverse trend of residential ESS, other types of ESS may enter residential market in one way or the other, such as Vanadium Redox Flow (VRF) ESS and Supercapacitor. Compared to LiB storage, VRF ESS and Supercapacitor have lower gravimetric energy density and corresponding bigger footprint, however, both VRF ESS and Supercapacitor can provide longer cyclability and more stable safe operation. The three types of ESS have different cycling speed, Supercapacitor has the highest speed, followed by LiB, the VRF has the slowest speed, as shown in **Table 1** [21].

Solid-sate battery (SSB) broke into our view since 2020. Different from LiB, the electrolyte in SSB is solid state. Based on different materials used in SSB Anode and Electrolyte, the SSB can be categorized as Li metal Sulfide SSB, Li metal Oxide SSB, and Anode free SSB [22]. Compared to LiB, the gravimetric energy density of SSB is about 1.7–2 times bigger, the volumetric energy density of SSB is about 1.1–1.6 times bigger, moreover, the SSB can operate under higher temperature of above 100°C [22]. SSB market size is predicted to be around USD 1645.6 million by 2030 [23]. The technology challenge for SSB is to combine both organic and inorganic solid electrolytes for obtaining both lower flammability and higher energy density [22]. Silicon Anode is also a technology trend due to its lower cost, non-flammability, and slower degradation. But Silicon Anode can bring in mechanical issues like voids and cracking [22]. In residential

	LiB storage	VRF ESS	Supercapacitor
Gravimetric energy density, Wh/kg	~150–250	~10–20	~1–10
C-rate	~0.2–3C	~0.1–0.3C	≥10C
Cyclability, years	~10–15	~20–30	~15–20
Therma runaway risk	Yes	No	No
Cost, USD $/kWh	~0.07–0.2	~0.05	~0.006

Table 1.
Performance comparison among different ESS solutions.

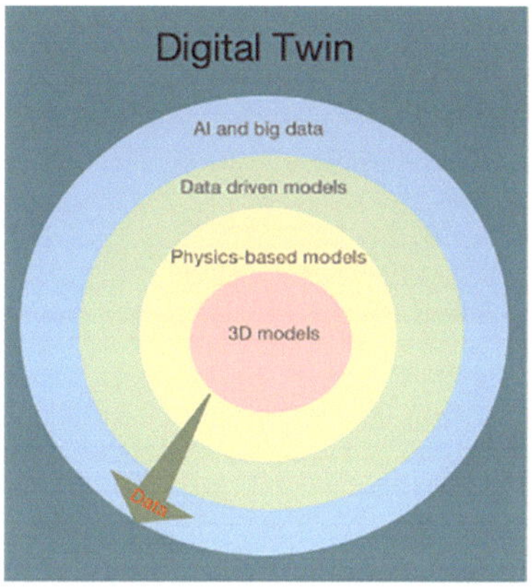

Figure 9.
An understanding of Digital Twin architecture in RRES.

market, there already exists SSB based products, this will motivate the development of residential ESS towards better safety, better reliability and smaller footprint.

Apart from different residential ESS solutions, the intelligent BMS will also dominate future technology trends by enabling predictive maintenance models in the cloud. The predictive maintenance model can generate adaptive warranty models and optimize quotation in product proposals. This will greatly sharpen company's business competency and enhance more opportunities.

Overall, a virtual system is anticipated to upgrade system performance, improve system operation efficiency and detect potential safety risks. It will be a bonus to have the virtual system alive to simulate different what-if scenarios. Thanks to all the research work published, we can now imagine such a virtual system architected with multi-discipline simulations and AI big data models using real-time data, as demonstrated in **Figure 9**. The virtual system also ages along with time and evolves to always mirror the physical system. In addition, the virtual system can have real-time communication with the physical system. This virtual system is widely recognized as 'Digital Twin.' We may not be able to develop fully working Digital Twin at this moment, however, some promising progress is already done by experts and colleagues in this field.

5. Conclusion

This chapter introduces the residential renewable energy solution (RRES) and the indispensable energy storage system (ESS) in RRES. The Li-ion battery (LiB) storage system, as the main focus, is introduced and analyzed with summarized methodology. The future development trend in RRES and residential ESS is discussed as well. I hope this chapter can help the people who are interested to learn renewable energy applications, as well as the people who want to push the existing boundaries in RRES.

Thanks

Special thanks to my family support, to Guan, Huiyi and Huiyou. Big thanks to Tamir Lance for his full-scale review and advice.

Author details

Ya LV
Innovation and Development in Renewable Energy Industry, Singapore

*Address all correspondence to: laylaneylv@gmail.com

References

[1] Available from: https://www.wsj.com/story/the-roots-of-renewable-energy-7993f651

[2] Available from: https://www.statista.com/statistics/873116/solar-energy-capacity-singapore/

[3] Available from: https://www.solarfeeds.com/mag/solar-power-statistics-in-china-2021/

[4] Available from: https://www.solarpowerportal.co.uk/blogs/uk_installed_730mw_of_solar_pv_in_2021

[5] Available from: https://taiyangnews.info/tag/eu-market-outlook-for-solar-power-2021-2025/

[6] Available from: https://www.seia.org/research-resources/solar-market-insight-report-2021-year-review

[7] Available from: https://www.pv-tech.org/india-deploys-8-4gw-of-solar-pv-in-h1-forecast-to-reach-20gw-by-end-of-the-year/

[8] Available from: https://www.greenmatch.co.uk/blog/2014/11/how-efficient-are-solar-panels

[9] Residential energy storage system market size, share & industry analysis, by technology (lithium-ion battery, lead acid battery, others), by application (on-grid, off-grid) and regional forecast 2022–2029. In: Market Research Report. Fortune Business Insight; 2022

[10] Available from: https://www.ema.gov.sg/statistic.aspx%3Fsta_sid%3D20140826Y84sgBebjwKV

[11] Available from: https://en.wikipedia.org/wiki/Lithium-ion_battery

[12] Available from: https://ulstandards.ul.com

[13] Available from: https://wwbridge-cert.com/certification/certification-by-certificate-type/iec

[14] Ahmad P, Shriram S, Gi-Heon K. Large Format Li-Ion Batteries for Vehicle. 2013. Large Format Li-Ion Batteries for Vehicle. Available from: http://www.nrel.gov/docs/fy13osti/58145.pdf

[15] Lam L. A Practical Circuit based Model for State of Health Estimation of Li-ion Battery Cells in Electric Vehicles. Master of Science Thesis, Faculty of Electrical Engineering, Mathematics and Computer Science Department of Electrical Sustainable Energy. University of Technology Delft; 2011

[16] Agwu DD, Opara FK, Chukwuchekwa N, Dike DO, Uzoechi L. Review of Comparative Battery Energy Storage Systems (Bess) for Energy Storage Applications in Tropical Environments. Owerri, Nigeria: Department of Electrical and Electronic Engineering, Federal University of Technology (FUTO);

[17] Ya LV. Semi-analytic method in fast evaluation of thermal management solution in energy storage system. International Journal of Energy and Environmental Engineering. 2020; **14**(11):349-353

[18] Available from: https://www.statista.com/statistics/1103401/predicted-lithium-ion-battery-capacity-by-company/

[19] Available from: https://cen.acs.org/materials/energy-storage/time-serious-recycling-lithium/97/i28

[20] Ya LV et al. Battery grading algorithm in 2nd-life repurposing li-ion battery system. International Journal

of Energy and Power Engineering. 2021; **15**(8):312-317

[21] Available from: https://www.nextbigfuture.com/2017/08/supercapacitors-game-changing-improvement-on-energy-density-compared-to-batteries.html

[22] Available from: https://www.battery.associates/battery-whitepapers

[23] Available from: https://www.precedenceresearch.com/solid-state-battery-market

Chapter 5

Balancing Renewable Energy Capacity, Time of Use Tariffs and Energy Storage in Energy Systems

David R. Walwyn

Abstract

The intermittency of solar energy predicates the simultaneous use of energy storage to maintain secure supplies. However, storage is expensive to instal and maintain, suggesting that there is an optimum design based on the price tolerance of electricity markets. In this chapter, a method for the calculation of the optimal size of a battery energy storage system (BESS), linked to utility-scale photovoltaic (PV) capacity, is presented. The method, which is illustrated by its application to the South African national grid (GridSA), uses historical generation/demand data to construct a spreadsheet model of the energy system. The model assumes that the difference between base load and energy demand, referred to as headroom, will be met using variable energy sources, including wind, solar, diesel/gas and batteries. Optimal sizing of these components to minimize the use of gas in summer, and make maximum use of low-cost solar and wind, leads to a configuration for GridSA consisting of a 22 GW base load (coal and nuclear), a PV installed capacity of 17.8 GW and a BESS capacity of 3.7 GW/10.4 GWh. A peak time of use tariff of ZAR3,500 per MWh (almost double the average tariff) will be optimal to build an economic case for energy storage as a sustainable option for GridSA.

Keywords: renewable energy, battery energy storage, time of use tariff, economic model, intermittency

1. Introduction

Intermittency is a term used to describe the variability of renewable energy sources such as solar, wind and wave. These sources are not consistently available, as they depend on specific weather conditions. Solar panels, for example, can only generate electricity when there is sufficient illumination and wind turbines need a minimum air velocity to turn. As a result, the output of renewable energy facilities can vary significantly over time. Demand for electricity, on the other hand, is continuous, although also variable in absolute quantities, with higher demand during the 'awake' day (05 h00 to 20 h00) and lower consumption during the 'asleep' night (20 h00 to 05 h00).

Typical profiles for wind and photovoltaic (PV) solar generation, and the demand for electricity by consumers in South Africa's national grid (GridSA) for the first week of December 2022, are shown in **Figure 1**. Solar is highly periodic and, to a large extent, predictable on a 5-day basis, but wind is highly variable and non-periodic, although the summer pattern often reflects strong winds in the late afternoon, which offsets the rapid decline in PV at sunset. Both characteristics of renewable energy generation are clearly visible in **Figure 1** and have also been previously reported [1].

The mismatch between supply and demand creates difficulties for grid management and has driven the development of large-scale energy storage systems, in addition to other approaches such as the increased deployment of gas and hydroelectric turbines. In the longer term, low-carbon storage solutions are essential if all countries are to reach net zero status by 2050. However, these systems are costly, and the cost of energy storage remains a significant barrier to the wider adoption of renewable energy.

In this chapter, the cost of battery-based storage, as revealed through a 2020 tender process in South Africa, is presented and discussed. Using real-time data for GridSA in June and December 2022, it is argued that the deployment of battery-based storage at times of high consumer demand is feasible but will require the wider implementation of time-of-use tariffs (ToUTs) in order to offset the additional cost. Estimates for the scale of the necessary systems, and the necessary ToUTs, are outlined. In all the discussions, GridSA is used as an illustrative example for the main tenets of the chapter. Section 2 presents the background information for the study, including a brief overview of GridSA, the use of renewable energy, the risk mitigation programme, the recent operation of the grid to manage the shortfall between demand and supply, referred to as 'loadshedding' and the practice of ToUTs. Section 3 describes the modelling of electricity supply and demand, and how this can be used to determine a system-level response to ToUTs. The proposed ToUT structure and levels to support greater deployment of battery electric storage systems (BESS) are then discussed in Section 4, followed by the study's conclusions in Section 5.

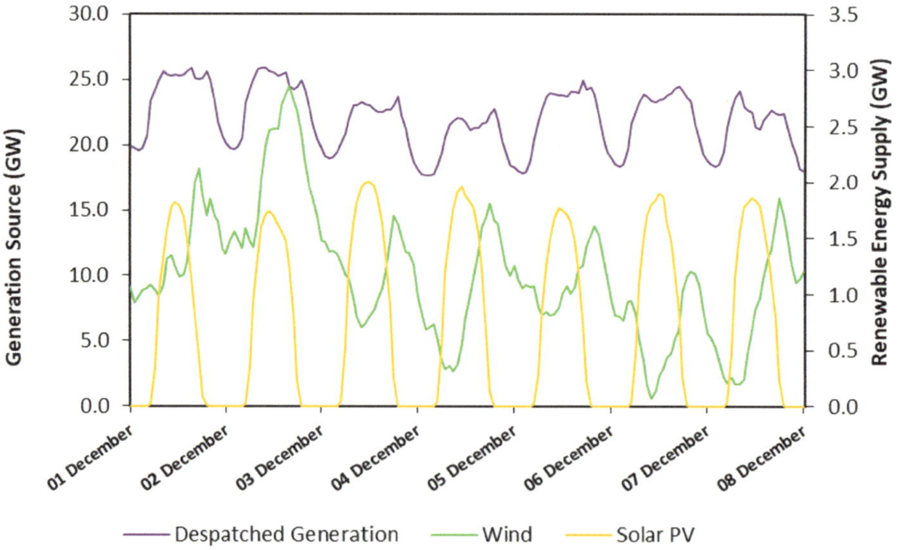

Figure 1.
Typical summer profiles for wind, PV and grid demand in South Africa.

DOI: http://dx.doi.org/10.5772/intechopen.111433

2. Background

2.1 Overview of the South African electricity grid

There are two striking features of GridSA, which shape the structure of the grid itself, its impact on the environment and its relationship with its customers. The first is the dominant role in all three aspects of electricity supply (generation, transmission and distribution) played by a single entity, namely the state-owned energy utility, Eskom, as shown in **Figure 2**. Eskom generates about 180 TWh per annum of electricity, supplying 90% of all electricity used in South Africa [2]. It operates a fleet of 15 coal-fired power stations, one nuclear reactor, 4 gas/diesel turbines and 7 hydroelectric schemes. Its fleet has a total generation capacity of 45 GW, but presently operates at a capacity utilisation factor of less than 50% [3]. The utility's transmission system includes about 28,000 km of high-voltage and 325,000 km of lower-voltage lines, and it supplies directly about 6.7 million direct customers [4].

The second feature is that GridSA is heavily dependent on coal as the main source of energy [5]. For example, as shown in **Figure 3** for the first week of December 2022, coal accounted for 80% of the total energy supplied by the utility. Other sources include diesel, PV, concentrated solar power, wind, hydro and nuclear [6]. Renewables account for 13% of the total supply with the remainder being obtained from nuclear (5%) and diesel (2%).

Demand in the grid follows a conventional square wave pattern, with a small evening peak linked to normal residential activities such as cooking, space cooling and water heating. Higher 'awake' day use is met with the combined use of non-dispatchable generation sources (wind and solar), diesel turbines and pumped storage. Users of electricity are similar to most industrialised countries with the main consumers being industry (51%), residential sector (20%) and the commercial/service sector (15%) [5]. Total demand is about 205,635 GWh, equivalent to an average power need of 23.5 GW.

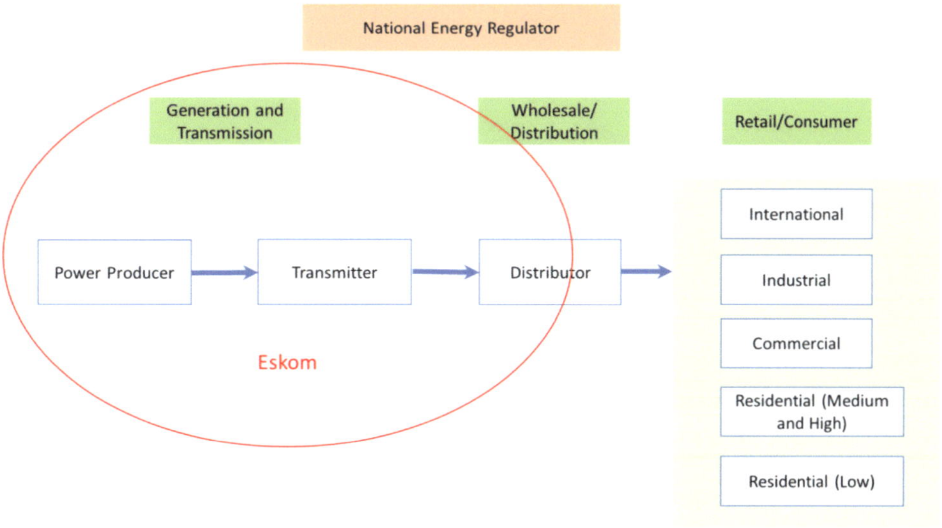

Figure 2.
Components of the electricity value chain.

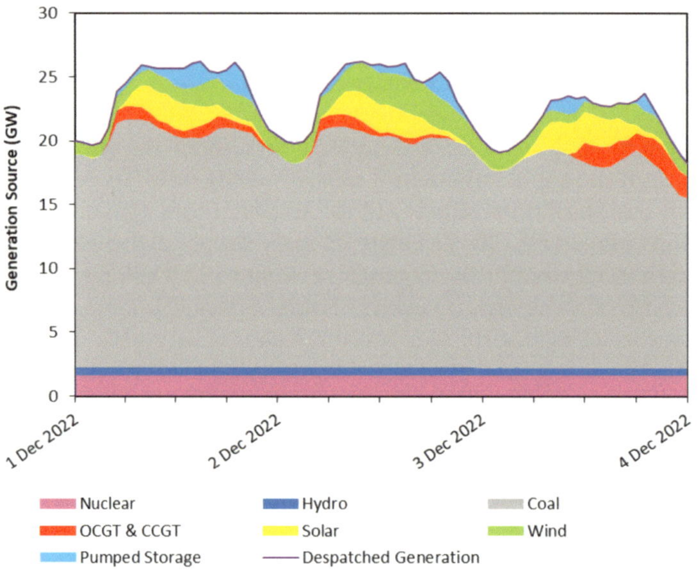

Figure 3.
Profile of GridSA energy supply (December 2022). Source: Own data and the Eskom data portal [3].

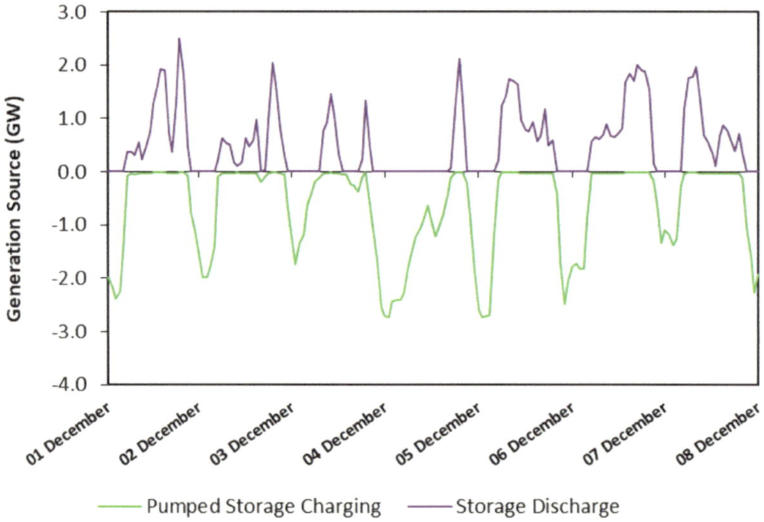

Figure 4.
Load and discharge cycles for GridSA's pumped storage (December 2022). Source: Own data and the Eskom data portal [3].

Pumped hydro is a key feature of GridSA and is a central to the later discussion in this chapter on loadshedding and load-shifting. Eskom operates three pumped storage schemes with a combined capacity of 2.73 GW. Excess energy generated during the night and over weekends by the coal-fired power stations is used to pump water from low to high altitudes; the same water is then returned to low-altitude storage dams during the day, as shown in **Figure 4**. The system has an overall efficiency of only 75% but it allows Eskom to flatten the demand curve and make full use of its available generation capacity during off-peak periods.

The high dependence on coal and pumped storage, which exists due to abundant and cheap coal deposits, and a specific investment programme for new power stations implemented in the 1960s to the 1980s, has become a significant barrier to the decarbonisation of the South African economy [7]. Although South Africa is a signatory of the 2015 Paris Agreement [8], it has made little progress towards the attainment of its nationally determined contributions [9, 10]. The rapid acceleration of its renewable energy programme remains essential if the country is to reach these targets. The programme, referred to as the Independent Power Producers Procurement Programme (REI4P), made some initial gains [11], but these have been largely side-lined and the country is now several years behind its energy transition programme, as outlined by the government's own policy papers [12].

2.2 Energy system reform and transition

Policy proposals to reform the electricity sector in South Africa have been tabled since at least 2008 with the introduction of the Electricity Pricing Policy [13]. The latter set out a number of policy objectives, including the liberalisation of energy markets, the introduction of a renewable energy programme, the establishment of a price control system to ensure that 'an efficient licensee can recover the full cost of its licensed activities' and support for energy access through affordable electricity tariffs [13].

Some of these objectives have since been realised, with the operationalisation of the National Energy Regulator of South Africa (NERSA) and the implementation of the REI4P [11]. However, in most respects, the policy proposals have been largely ignored and the utility has sharply regressed since the 2000s. Electricity prices have risen sharply and are 750% higher than in 2008, increasing at a rate 3 times faster than the rate of inflation. Eskom's debt has ballooned to R400 billion, and its interest expenses exceed operating profits by at least R3 billion, making the utility technically insolvent. Its environmental footprint remains unsustainable, and the utility is now the largest sulphur dioxide emitter in the world, exceeding the total emissions of China and the United States combined. Its carbon emissions are about 200 million tonnes per year, representing 40% of South Africa's total emissions, and all of Eskom's 15 power stations are in breach of the minimum emission standard (MES). The renewable energy programme, which could have partly rescued the country from its energy crisis and environmental non-compliance, was put on ice over the critical period of 2014 to 2021 and is still at least 5 years behind schedule, with a deficit of about 12 GW of renewable energy capacity relative to the initial Integrated Resource Plan 2010 [12]. Finally, the liberalisation of energy markets, including the unbundling of Eskom and the formation of regional electricity distributors, has never materialised.

Other studies have similarly noted a lack of progress towards the policy objectives [14, 15]. One of the more severe consequences of this failure is the growing crisis of supply constraints, the details of which are covered in the following section.

2.3 System constraints and loadshedding

The country has a history of energy shortages, including electricity blackouts in 2007 and 2008, petroleum shortages in 2008 and 2011 and gas shortages in 2011 and 2012 [16]. More recently, Eskom has been unable to meet electricity demand over long periods and has implemented a programme of loadshedding or rolling blackouts, in which power to customers is interrupted on a rotational basis, depending on the level of energy savings to be realised.

The system is more easily understood using the diagram in **Figure 5**. In the first week of December, Grid SA experienced persistent energy shortfalls, leading to progressive stages of loadshedding. For the first 2 days, supply and demand were largely balanced except overnight on the 1st of December, resulting in the use of Stage 1 loadshedding. The notation of 1 vs. 2 vs. 3, etc. refers to the shortfall between anticipated demand and actual supply. Figures for the former are obtained by Eskom from historical data for energy demand, adjusted for both seasonal and daily fluctuations. Stage 1 loadshedding, therefore, refers to a gap of 1 GW between demand and supply, Stage 2 to a gap of 2 GW and so on.

As the supply crisis deepened during the first week of December 2022, the levels of loadshedding increased, as shown in **Figure 5**. By the end of the week, the power shortfall was 4 GW, resulting in Stage 4 loadshedding.

The impact of loadshedding on electricity consumption can also be deduced from **Figure 5**. Although the average demand is 27 GW, there are a group of customers, known as the strategic facilities, where the power is always maintained. Load-shed customers constitute about 18 GW or 67% of the total market. In the event of a supply shortfall of, for example, 2 GW, non-strategic customers lose power for 2.7 (= 2/18*24) hours, making the duration and frequency of loadshedding more severe than might be the case if the blackouts were spread over the whole customer base.

This discussion of system constraints and the practice of loadshedding is germane to the remainder of this chapter, whose focus is whether ToUTs can be used to support the initial introduction of battery energy storage with high levels of renewable energy, particularly PV, within Grid SA. The pattern of loadshedding provides some insight into the question of possible consumer response to ToUTs. This topic is further discussed in Section 4. In the next section, the rationale for a ToUT is explained.

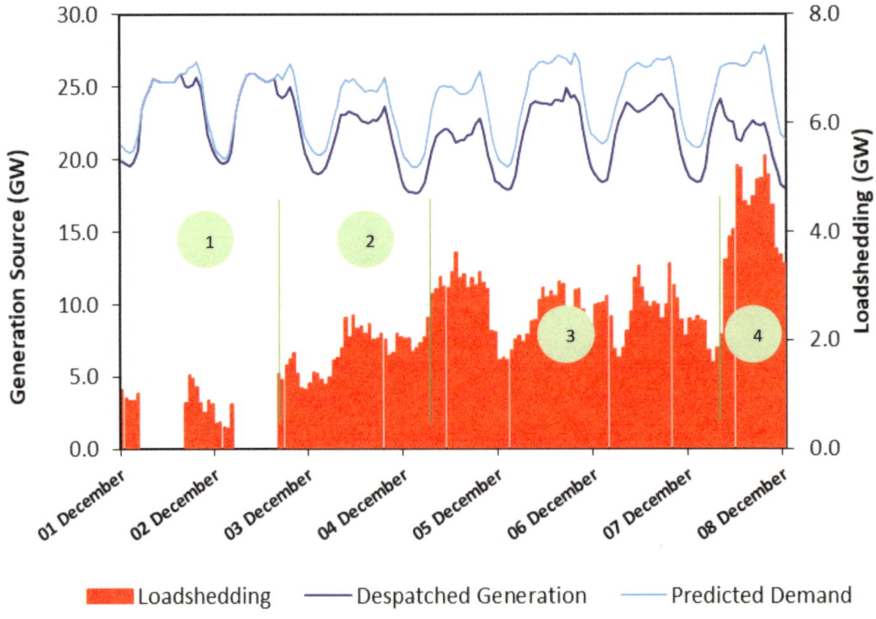

Figure 5.
Loadshedding and energy supply in the GridSA.

2.4 Time of use tariffs

ToUTs are pricing structures for electricity that vary the price depending on the time of day and the season. They are designed to incentivise consumers to use electricity when it is more plentiful or cheaper to produce, thereby reducing their usage during times when the demand or the cost of production is high. ToUTs are one example of several approaches employed by utility companies for demand-side management, with other examples, including the introduction of mechanisms to encourage energy efficiency and loadshedding during periods of high demand.

There are typically three pricing periods for ToUTs: peak, off-peak and shoulder. Peak periods are when demand for electricity is highest, and the price is highest. Off-peak periods occur when demand is lowest, and the price is lowest, mostly during the night. Shoulder periods take place between peak and off-peak, and the price is generally lower than the peak period but higher than the off-peak period.

ToUTs can have positive impacts, for example acting as a tool to manage electricity demand and helping to balance supply/demand in the grid. They can also help to reduce the need for gas- or diesel-based peaking power plants that are only used to meet peak demand and have high environmental impact; they can improve system reliability; they can reduce overall energy costs of consumers and they can lower the investment costs for utility companies.

However, there are some limitations to the use of ToUTs as mechanisms to change consumer behaviour. For instance, in a pilot study of Irish households with smart metering and ToUTs, it was found that instantaneous feedback on energy savings was essential in developing sustained changes to usage, particularly a shift in demand away from peak periods, but a steeper tariff profile (between off-peak and peak rates) had little further impact [17]. Consumers with higher levels of education and income were also more responsive to ToUTs [17].

Notwithstanding the limitations, ToUTs or real-time pricing will be essential to the stable operation of Grid SA, indeed any national grid, with high levels of renewable energy generation, and especially high levels of PV [18]. In the absence of effective demand side management, systems will suffer from the well-known 'duck curve' effect, characterised by an oversupply of energy at midday as solar reaches its peak, and a sudden undersupply in the evening as the sun sets and demand accelerates to the evening peak [19, 20].

It is noted that Eskom already applies ToUTs, as listed in the tariff schedule [21]. The off-peak and low-season tariffs are about 50% of the peak values, providing a large incentive for consumers to adjust their energy practices. However, the benefit of lower tariffs can only be realised if consumption is managed through smart metering, connected to an appropriate billing system. Most households and low-energy users do not use such systems, and hence cannot take advantage of the significant discounts presented by the Eskom ToUTs.

In summary, ToUTs, supported by smart metering and changes to consumer behaviour, are critical to higher levels of renewable energy generation in Grid SA. The remaining question for this chapter is the estimated level of the necessary ToUT, which would drive behavioural change and provide a reasonable return on investment to the utility company. This question is now covered.

3. Modelling the energy system

Modelling of GridSA was undertaken with spreadsheet models using real-time data downloaded from the Eskom Data Portal [3]. The raw data were first processed to generate profiles for total demand, and energy production per source of supply. Simulations of the required energy storage, and hence the necessary tariff, were then developed for a winter month (June) and a summer month (December).

The accuracy of the modelling depends, inevitably, on the quality of the data from the portal. Assessments of the extent of power generation outside GridSA, unrecorded on the data portal, are that this is limited to about 5% of the total energy. In other words, 95% of electricity supply is covered by the data portal.

In order to model the required tariff, data from the Risk Mitigation Independent Power Producers Procurement Programme (RMI4P) were used as the benchmark for the cost of providing energy in the PV shoulder periods (early morning and late evening). The supply crisis, as outlined in Section 2.3, precipitated the initial conceptualisation of this programme by the South African government in 2020 [1]. The main objectives of the programme were to establish independent power producers able to provide 2 GW of emergency power on a flexible basis between the hours of 05 h00 and 21 h30, and to be able to respond to needs of the electricity system operator based on an automatic generation control load-following ability [22].

Following the call for proposals issued in August 2020, 28 bids were received by the South African government, of which 7 were announced as preferred bidders in mid-2021. The weighted average feed-in tariff for the winning bids was ZAR1.61/kWh[1], with the sum of the bid capacity being 2 GW. A summary of the preferred renewable energy-based bids is given in **Table 1** (one bid based on natural gas has been excluded).

The bids provide a clear indication of the cost of electricity over the peak period, relative to the Eskom standard tariff and the present price of solar. Benchmarks for the former values can be obtained from the Eskom booklet on tariffs and schedules [21], and for the latter from the Bid Window 6 contracts, recently awarded to

Bidder	Size (MW)	Energy Cost (ZAR/kWh)	Gas/Diesel (MW)	PV (MW)	Wind (MW)	BESS	
						MW	MWh
ACWA	150	1.46	15	422	0	150	900
Mulilo/Total Coega	198	1.89	198	216	0	0	0
Mulilo/Total Hydra	75	1.52	20	216	0	150	600
Omoyilanga	75	1.72	12	138	77	75	450
Oya	128	1.55	106	155	83	40	400
Scatec	150	1.88	0	150	0	540	2250

Table 1.
Summary of preferred RMI4P bids.

[1] The exchange rate (March 2023) is ZAR18.4/USD

independent power producers [23]. The values are ZAR1,250 per MWh and ZAR490 per MWh, respectively.

The real-time data and the values from the RMI4P provided the basis for the calculation of energy tariffs, the results of which follow in the next section.

4. Proposed tariff structure to support battery storage

The central question of this chapter is whether ToUTs can be used to allow greater penetration of PV within GridSA. There are two reasons for this proposition, the first being that ToUTs could prevent the occurrence of the notorious duck curve, and the second being that the additional revenue from a peak tariff and/or shoulder tariffs could be used to justify the use of BESS.

As indicated earlier, the loadshedding data presented in Section 2.3 and **Figure 5** provide some insight into possible consumer responses to ToUTs. Interestingly both the total predicted, and the actual curtailed, demand have similar profiles, indicating little change to hourly consumption. Electricity consumption appears to be mostly instantaneous or inflexible; when supply is curtailed, consumers cannot power their homes, devices or activities from the grid, and either do not use energy or find alternative means of supply, such as local diesel-based generators or PV. The similar pattern of the two curves in **Figure 5** suggests that there is limited load-shifting in response to loadshedding, or use of BESS. In other words, further load shifting, other than that already introduced by the Eskom ToUTs, may not be realised with ToUTs; instead, customers will use the available electricity, despite its higher cost. This behaviour reflects supply, and not demand, elasticity.

This anticipated response provides a strong basis for the implementation of BESS, despite its additional cost to Grid SA, as long as this cost remains competitive to independent power generation already being accessed by customers on the grid, such as PV and diesel. Indicative values obtained from the RMI4P tender prices suggest that this requirement will indeed be possible. For instance, the price from Mulilo/Total Hydra proposal of **Table 1** for emergency power in the shoulder and peak periods is ZAR1,520 per MWh (vs. a diesel-based tariff of ZAR6,000 per MWh).

Mulilo/Total Hydra is a particularly relevant model for the issue being explored in this chapter. The proposal consists of three generation technologies, namely PV (216 MW), BESS and a small quantity of diesel (20 MW). The project was designed to supply 70 MW of power to the grid, as shown in **Figure 6**. The profiles indicate that solar component was overdesigned to ensure sufficient power for recharging the BESS under both winter and summer conditions. During the latter, there is no need for gas-based generation, but in winter, the lower peak of solar-based generation and the increased likelihood of cloudy skies requires use of gas/diesel on certain days (see 9th June in **Figure 6**).

The data from Mulilo/Hydra can now be extended to the broader question of whether a PV/BESS configuration could meet the electricity demand of Grid SA, and, if so, at what additional cost. The first assumption in this analysis is to extend the concept of headroom, as previously developed for the national grid in the United Kingdom [24], to GridSA. The concept considers that supply can be separated into two components, namely the base load and the headroom, where the latter is the difference at any one moment in time between total demand and base load, as given in Eq. 1.

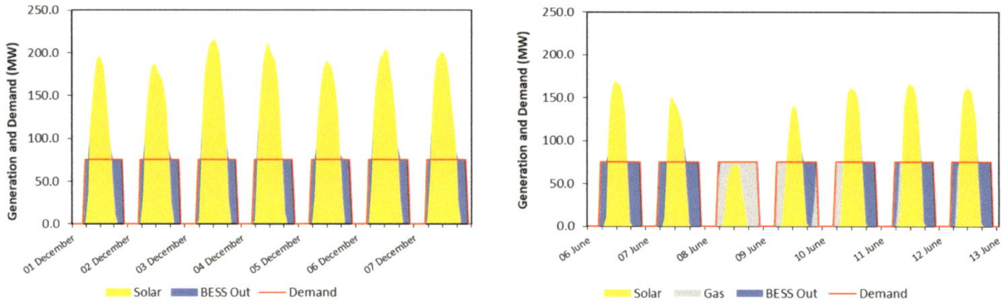

Figure 6.
Energy generation (by source) and supply for Mulilo/Total hydra (summer and winter).

$$\text{Total Energy Supplied} = \text{Base Load} + \text{Headrooom} \tag{1}$$

Base load on GridSA is typically about 22 GW, with remainder being supplied by wind, pumped storage, solar and diesel/gas turbines, all of which are variable or intermittent inputs. Adopting the value of 22 GW in the model, and assuming a wind energy input of 5 GW, which is 175% of the present capacity on GridSA, it is now possible to calculate the required PV capacity from Eq. 2, the historical data for the summer and winter months and a summer PV efficiency in South Africa of 37%.

$$\text{Required } PV = \frac{\max(\text{Headroom} - \text{Wind})}{\text{PVEfficiency}} \tag{2}$$

For the 2022 data set, which was used as the basis of this study, the summer and winter values for the largest difference between the headroom and the energy derived from wind are calculated to be 4.4 and 8.3 GW, respectively, giving a required PV capacity of 17.8 GW (summer conditions and a 50% overdesign factor). On the basis that the summer demand should be met by wind and solar only, without the use of gas turbines, the BESS power rating and the total energy storage can now be calculated using the following algorithm:

1. Input the PV capacity and calculate the maximum BESS output for the summer conditions.

2. Enter this value in the spreadsheet as the design power capacity for the BESS facility.

3. Calculate the necessary BESS energy rating such that in summer conditions, the gas turbines are not required.

This approach leads to the result that the optimal BESS specification is 3.7 GW/10.4 GWh. The output of this proposed system is shown in **Figure 7**. It is noted that 10.4 GWh is significantly larger than the present BESS facilities; the largest existing BESS is the Moss Landing Energy Storage Facility in California with a capacity of 1.6 GWh. Multiple facilities giving a total storage capacity of 10.4 GWh over several sites are, therefore, recommended.

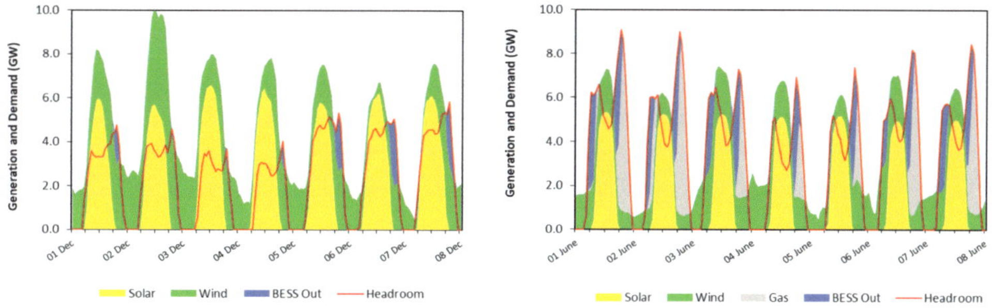

Figure 7.
Proposed solar/BESS energy generation for grid SA (winter and summer).

The cost of such an arrangement, and the level of ToUTs to support the investment, must now be considered. The bid price for the Mulilo/Total Hydra proposal of the RMI4P was ZAR1,520 per MWh. Using the spreadsheet model of Mulilo/Total Hydra, it is calculated that BESS will contribute an average of 23% to the total energy output. Moreover, it has already been mentioned that the current cost of PV is ZAR490 per MWh, leading to the result that the BESS contribution to the total cost is ZAR4,970 per MWh. It is noted that this estimate is similar to other values reported in the literature of ZAR4,000 to 5000 per MWh [25–27].

The same calculation can be repeated for the GridSA model. Given that wind and solar capacity are overdesigned and, as shown in **Figure 7**, this approach leads to significant excess generation, it is also assumed that the additional energy can be absorbed by GridSA's pumped storage capacity, hence maintaining a low unit energy cost. In the extreme case, such as may occur in winter, the evening peak period would be supplied by wind, diesel and batteries in the proportions of 30, 40, and 30%, respectively, from which the required evening tariff can be calculated at ZAR3,500 per MWh.

It is also evident from **Figure 7** that PV/BESS combination introduces another factor to the discussion on ToUTs. Peak power during daylight hours can be met reliably with solar PV, but shoulder energy demand, which lies typically between 06 h00 and 09 h00 in the mornings and 16 h00 to 19 h00 in the evenings, must be supplied by either BESS or diesel. The latter is a more expensive option and should, therefore, be allocated a higher tariff. The earlier discussion on cost suggests that this tariff could be set at nearly double the peak tariff, the latter being presently about ZAR2,000 per MWh in South Africa.

Patterns of consumption during periods of loadshedding indicated that consumers adopt a behaviour of supply elasticity, preferring to find alternative sources of energy during loadshedding rather than to shift consumption to other times of the day. This response, as already noted, is not surprising. Most entities, including schools, commercial businesses, hotels, restaurants and retail outlets, have fixed hours of work with little flexibility in terms of their operations. Finding alternative sources of electricity is then not a matter of choice, it is a necessity.

At present, much of the independent power production relies on small-scale diesel, although PV is growing throughout the commercial and residential sectors. The cost of these alternatives is close to ZAR6,000 per MWh, suggesting that a Grid SA shoulder tariff of ZAR3,500 per MWh may be 'accepted' by the market, particularly if the cost of peak power (used over the period 09 h00 to 16 h00) could be reduced through an increase use of PV. The latter has a levelised cost of energy, which is nearly one-quarter of the present retail price for electricity in South Africa, as already noted in the earlier discussion.

5. Conclusion

The advent of renewable energy, particularly PV, heralds the possibility of widespread low-cost energy with a vastly reduced environmental footprint [11]. Solar conditions in South Africa are excellent for energy generation, and when combined with energy storage systems such as pumped storage or BESS, could lead to the necessary reform of Grid SA.

However, BESS is costly to instal and presently not feasible at the scale, which would be required to support energy demand within South Africa. In this chapter, the use of ToUTs has been explored as a means of providing an economic incentive for private investors, or the state, to build large-scale BESS, coupled with PV. Using a time series model based on real-time data, and information collected from the RMI4P, it is calculated that 6.3 GW of effective capacity would be optimal for a PV/BESS facility, which could be met under summer conditions with a PV installed capacity of 17.8 GW and a BESS capacity of 3.7 GW/10.4 GWh.

The facility would require a ToUT during the shoulder periods of ZAR3,500 per MWh, which is nearly double the present Eskom tariff and will lead, at least initially, to consumer resistance. However, data from the periods of loadshedding in South Africa reflect supply elasticity in the behaviour of South African consumers. When supply is curtailed, consumers either survive without power, or they access other sources of electricity, even if the latter is more costly, such as diesel and rooftop solar. Furthermore, the shoulder ToUT could be offset by a lower peak tariff, made possible through the widespread use of PV with a levelised cost of energy of about one-third of the Eskom tariff.

This pattern of behaviour suggests that the introduction of ToUTs with smart metering will enable energy companies and Eskom to recover the additional cost of BESS, which will be necessary to support the use of higher levels of PV within Grid SA.

Author details

David R. Walwyn
Department of Engineering and Technology Management, University of Pretoria, Pretoria, South Africa

*Address all correspondence to: david.walwyn@up.ac.za

References

[1] Walwyn DR. Providing emergency power in national energy systems: South Africa's Risk Mitigation Programme is costly and environmentally disastrous. In: 2021 International Conference on Electrical, Computer and Energy Technologies (ICECET). Cape Town: IEEE; 2021. pp. 1-5

[2] Eskom. Integrated Annual Report 2021/22. Johannesburg: Eskom; 2022

[3] Eskom. Eskom Data Portal. Johannesburg: Eskom; 2021

[4] Eskom. Company Information Overview: Eskom. 2021. Available from: https://www.eskom.co.za/about-eskom/company-information/

[5] International Energy Agency. Electricity generation by source. 2021. Available from: https://www.iea.org/regions/

[6] Pierce W, Ferreira B. Statistics of Utility-Scale Power Generation in South Africa in 2021. Pretoria: CSIR; 2022

[7] Parr B, Swilling M, Henry D. The Paris Agreement and South Africa's Just Transition. Melbourne: Melbourne Sustainable Society Institute; 2018

[8] UNFCCC. FCCC/CP/2015/L.9/rev.1: Adoption of the Paris Agreement. Paris, France: UNFCC; 2015

[9] Walwyn DR. Turning points for sustainability transitions: Institutional destabilization, public finance and the techno-economic dynamics of decarbonization in South Africa. Energy Research & Social Science. 2020;**70**:101784

[10] Tyler E, Hochstetler K. Institutionalising decarbonisation in South Africa: Navigating climate mitigation and socio-economic transformation. Environmental Politics. 2021;**30**(suppl. 1):184-205

[11] Walwyn DR, Brent AC. Renewable energy gathers steam in South Africa. Renewable and Sustainable Energy Reviews. 2015;**41**(1):390-401

[12] Department of Energy. Integrated resource plan 2019. In: Energy Do. Pretoria: Department of Energy; 2019

[13] Department of Minerals and Energy, editor. Electricity Pricing Policy of the South African Electricity Supply Industry. In: Department of Minerals and Energy. Pretoria: South African Government; 2008

[14] Todd I, McCauley D. Assessing policy barriers to the energy transition in South Africa. Energy Policy. 2021;**158**:112529.

[15] Presidential Climate Commission. Laying the Foundation for a Just Transition Framework for South Africa. Pretoria: Presidential Climate Commission; 2021

[16] Sparks D, Madhlopa A, Keen S, Moorlach M, Dane A, Krog P, et al. Renewable energy choices and their water requirements in South Africa. Journal of Energy in Southern Africa. 2014;**25**(4):80-92

[17] Di Cosmo V, Lyons S, Nolan A. Estimating the impact of time-of-use pricing on Irish electricity demand. The Energy Journal. 2014;**35**(2):119-138

[18] Sarfarazi S, Mohammadi S, Khastieva D, Hesamzadeh MR, Bertsch V, Bunn D. An optimal real-time pricing strategy for aggregating distributed

generation and battery storage systems in energy communities: A stochastic bilevel optimization approach. International Journal of Electrical Power & Energy Systems. 2023;**147**:108770

[19] Krietemeyer B, Dedrick J, Sabaghian E, Rakha T. Managing the duck curve: Energy culture and participation in local energy management programs in the United States. Energy Research & Social Science. 2021;**79**:102055

[20] California Independent System Operator. Fast Facts: What the Duck Curve Tells Us About Managing a Green Grid California 2013. Available from: http://www.caiso.com/documents/flexibleresourceshelprenewables_fastfacts.pdf

[21] Eskom. Tariffs & Charges Booklet 2022/2023. Johannesburg: Eskom; 2022

[22] IPP Office. IPP Risk Mitigation. Pretoria: IPP Office; 2021

[23] IPP Office. List of Preferred Bidders for Bid Window 6 as Announced on 8 December 2022. Pretoria: IPP Office; 2022

[24] Stephens A, Walwyn DR. Wind energy in the United Kingdom: Modelling the effect of increases in installed capacity on generation efficiencies. Renewable Energy Focus. 2018;**2018**(27):44-58

[25] Mayyas A, Chadly AA, Khaleel I, Maalouf M. Techno-economic analysis of the Li-ion batteries and reversible fuel cells as energy-storage systems used in green and energy-efficient buildings. Clean Energy. 2021;**5**(2):273-287

[26] Lazard. Levelized Cost of Storage. New York: Lazard; 2020

[27] Roy S, Sinha P, Ismat SS. Assessing the techno-economics and environmental attributes of utility-scale PV with battery energy storage systems (PVS) compared to conventional gas peakers for providing firm capacity in California. Energies. 2020;**13**(2):488

Chapter 6

Prospection of Neighborhood Megawatthours Scale Closed Loop Pumped Hydro Storage Potential

Kiswendsida Elias Ouedraogo, Pinar Oguz Ekim and Erhan Demirok

Abstract

Energy is at the center of the global socio-economical, geopolitical, and climate crisis. For this reason, countries are looking to boost their energy independence through the integration of distributed green electricity. However, the bottleneck of intermittent renewable energy as an alternative to fossil fuel energy remains the high cost of large-scale energy storage. The study explored the existence of megawatt-hours scale closed-loop pumped hydro-storage reservoirs near communities. A MATLAB algorithm has been developed to detect 1, 4, 9 hectares reservoirs with a separation distance less than 1000 meters, and a head over 100 meters, corresponding to an energy capacity of 20 to 400 megawatt-hours per pairs. For the cities studied (Banfora, Syracuse, Manisa), the results revealed the existence of more than 10.000 megawatt-hours storage capacity in each city, which exceed the need of the communities. In the 4 hectares sites category, all cities have over 80 pairs of reservoirs ideal for distributed storage system implementation. Therefore, a 100% renewable energy power grid that is resilient, reliable, can be achieved faster by adopting distributed closed-loop pumped hydro-storage, which has limited environmental impact and is likely to attract a large number of smaller investors.

Keywords: pumped hydro storage, energy resilience, utility storage, long duration battery, community energy

1. Introduction

There is a worldwide urgency to solve the energy crisis sustainably in order to reverse climate change-induced socio-economic problems. To limit carbon emissions, renewable energy (RE) such as solar and wind power have been promoted by governments through subsidies in the last two decades. Nowadays, in many countries, RE became competitive even without subsidies. However, as intermittent RE penetration ratio increases, it also becomes mandatory to add energy storage system for grid stability, and to make the system a full alternative to fossil fuel power plants. The research on storage systems often categorizes the technologies based on the duration of usage, (long-short duration), and the discharge rate or power capacity [1]. However, it is not uncommon to see others performance indicators such as efficiency, cost, footprint, environmental friendliness, and so on be used [2]. In the short duration, high power density categories, there are chemical batteries such as lithium-ion batteries, there are flywheels, super-capacitors, and magnetic coils. In the long duration, low power density categories, there are pumped hydro storage (PHS), compressed air energy storage (CAES), and hydrogen storage [3]. For the energy transition purposes, the interest is in technologies capable of providing bulk storage cost-effectively in addition to scalable grid-level power output. In the literature, PHS and CAES appear to be the leading technologies regarding these criteria [2]. When efficiency is added into consideration, PHS is the net winner as its round efficiency is around 80% versus 40–60% for CAES [4, 5]. PHS has the benefit of loads shifting, avoiding RE curtailment, allowing black start of the grid, and transforming solar/wind plants into a fully dispatchable power unit [6]. Also, the mass of the turbine and the generator provides a natural inertia which contributes to grid stability. Given these characteristics, PHS capacity is projected to double by 2050 from 168 GW to over 300 GW. To revive the interest in hydropower, studies have been conducted to estimate the potential of small ($<$ 10 MW) hydropower sites worldwide [7]. It concluded with findings revealing that the installed capacity is far less than the non-used potential. In Africa, there are 595 MW installed plants and over 10.000 MW potential plants. In America, the ratio is 6 GW / 41 GW. In recent years, the factors affecting the development of hydro plants have been subject to studies [8]. Among the driving factors can be cited the need for more RE integration, revenue generation, support for rural development, and energy security. The barriers are mostly the high capital expenditure, the lack of accompanying infrastructure (roads, power lines), the lack of suitable land, and environmental issues. In fact, depending on the source, the cost of hydro plants varies widely from 600 $/kW up to 8000 $/kW [8, 9]. This is much above the cost of thermal plants and solar plants which are around 700 $/kW, and 1000 $/kW respectively [10, 11]. However, on the levelized cost of energy side, hydro plant is by far the cheapest option since no fuel is used and the plant can run 24/7. To reduce the cost for storage-only applications, closed-loop systems which do not require continuous water flow from a river are becoming more and more popular in the research. Usually, site exploration is done for suitable pair of upper reservoir and lower reservoir [12, 13]. Such a pair of sites have been already mapped by several studies including Australia global pumped hydro storage atlas which unveiled over half million potential pairs, totaling more than 20-million-gigawatt hours capacity [13]. In the United States, similar study uncovered the existence of more than 10 thousand suitable pairs of reservoirs mainly in the west coast [14]. However, the development of these giga scale infrastructures is capital intensive and is far from being an unanimous solution among decision markers. Therefore, it is necessary to push further the research to make the technology more acceptable by the society.

1.1 Objectives

In the literature, pumped hydro storage is often associated to large giga-scale projects. Due to these sizes, projects environmental risks and safety risks are so high and the implementation from permitting, construction to operation is a lengthy process that can take up to a decade [15]. Also, projects sites are often in remote areas, requiring expensive high power capacity transmission infrastructure. To mitigate the drawbacks of giga scale PHS, this study aims to:

- Assess the existence of lower capacity megawatt-hours scale PHS sites.
- Search for PHS site only in the neighborhood of communities.
- Fill the gap by prospecting 100 m – 300 m head sites.
- This choice is expected to have the following benefits:
- Integrate more easily distributed energy resources such as community solar PV.
- Reduce projects environmental risks, and implementation timeline.
- Attract more smaller investors to boost projects number faster.

2. Grid scale long duration energy storage technologies

There are much research efforts being conducted worldwide by both academics and industries to provide grid scale energy storage solutions. Technologies are competing for application specific technical suitability and costs over benefits ratio. These observations imply that, the future of energy storage is likely to be populated with a mix of solutions complementing each other.

2.1 Existing technologies

There is no consensus on the storage size that qualify as grid scale neither on the storage duration that qualify as long duration. However, in many countries, solar PV plant over 5 MW or hydro power plant over 10 MW are registered as utility scale facilities. On the storage sides, several projects with less than 4 hours have been tagged as short duration storage. In the context of this paper, grid scale, long duration storage are technologies capable of delivering MWs of power for a period from 10 hours to several days. A 1 MW, 20 MWh energy storage module is considered as a unit block to build grid-scale long duration storage. This is done as it was observed that modularity is one of the biggest factors that contributed to the success of solar or wind power in the market. With reduced unit capital expenditure (CAPEX), the system can be standardized, and built-in factory to reduce cost. Also, a wider range of investors small or big can be reached which is not the case for giga scale system where only a handful of international institutions and governments are the soles actors. Grid scale storage solutions can be grouped into 3 categories: electrochemical, thermal and mechanical. In the first group, there are lead acid batteries, lithium batteries, hydrogen fuel cell, redox flow batteries. Due to their small footprint, they have been deployed at strategic nodes of the grid to provide ancillary services mainly. In the second group, molten salt is the most popular technology. As phase change materials offer high latent heat at nearly constant temperature, they are ideal energy source for steam to electric systems. In Morocco for example, in 2019 a utility scale thermal storage has been combined to concentrated solar thermal to build a dispatchable electric power plant. In the last group, CAES and PHS can be cited. While CAES recorded few projects, PHS is up to date the largest contributor of the world storage capacity with over 90% of the share. To build on the success of PHS, other form of gravity energy storage using a denser fluid or solid blocks are being explored. Before going into details about gravity-based energy storage, it is worth reviewing the performance of different technologies.

2.2 Performance comparison

The typical power rating and discharge duration of some storage technologies are displayed in **Figure 1a**. PHS, CAES, flow battery, and hydrogen storage are the most promising for grid-scale long duration energy storage. A common point of these technologies is their ability to decouple energy capacity and power rating, which provide greater flexibility for system planning. In the contrary lithium batteries, nickel metal batteries or lead acid batteries have their power and energy linked by the cell chemistry. Often, high energy capacity batteries come with an unnecessary high-power rating. However, this situation makes them an excellent technology for grid frequency and peak regulation which need high burst of power in the seconds to hours time scale.

From an efficiency and lifespan perspective, **Figure 1b** shows that PHS is the optimum technology with a round trip efficiency of 80% and an extremely long life over a century, assuming a daily charge discharge. These characteristics make PHS an ideal solution to combine with solar PV or wind power plant which can last 30 years. On the lower left corner, lead acid and nickel metal batteries are the worst solutions because of the low number of cycles at 80% deep of discharge (DoD). From this fact, it can be concluded that they are not suitable for applications that require daily cycling. However, for emergency backup power need for example, they can be a good solution. Lithium batteries efficiency and number of cycles are well improved compared to lead acid batteries. For mobile applications such as electric vehicle, they are the current optimum solution. For grid scale storage, pilot projects of lithium battery are being rolled worldwide. As the lifecycle is limited to 10 years for 80% DoD, combining them with solar or wind power require 2–3 replacement per project life. This can represent a serious bottleneck that suppress the high efficiency benefit of lithium battery. Lastly for hydrogen fuel cell and CAES, they both suffer from a low round trip efficiency of 40%. However future development in fuel cells and electrolyzers efficiencies may improve hydrogen storage solution while heat recovery can boost the overall efficiency of CAES systems.

In terms of cost performance, **Figure 1c** shows that PHS and CAES are the best solution for grid scale application although CAES is much less widespread than PHS. From surveyed manufacturers, the 2022 cost of lithium battery is 300–500 $/kWh, that of flow battery 500–700 $/kWh which are much over the 50–100 $/kWh for PHS [19]. Even if chemical batteries cost drops to 100 $/kWh in the future, the rarefication of minerals, mining environmental cost, geopolitical instability, logistics costs, and other costs are likely to make PHS the preferred option for countries. In addition, the ongoing electrification of transportation will certainly divert an important portion of chemical batteries into mobile applications which have more value than grid storage stationary applications. By considering together discharge duration, efficiency, and cost performance, PHS outrank both CAES and chemical batteries for long duration grid scale applications. The biggest drawback of PHS is the dependency of terrain so research is focused on quantifying the existing exploitable potential.

3. Gravity based energy storage

Although water potential energy storage is the most used gravity-based storage (GES), the system can use other liquids, solid and even gas to work. Many GES

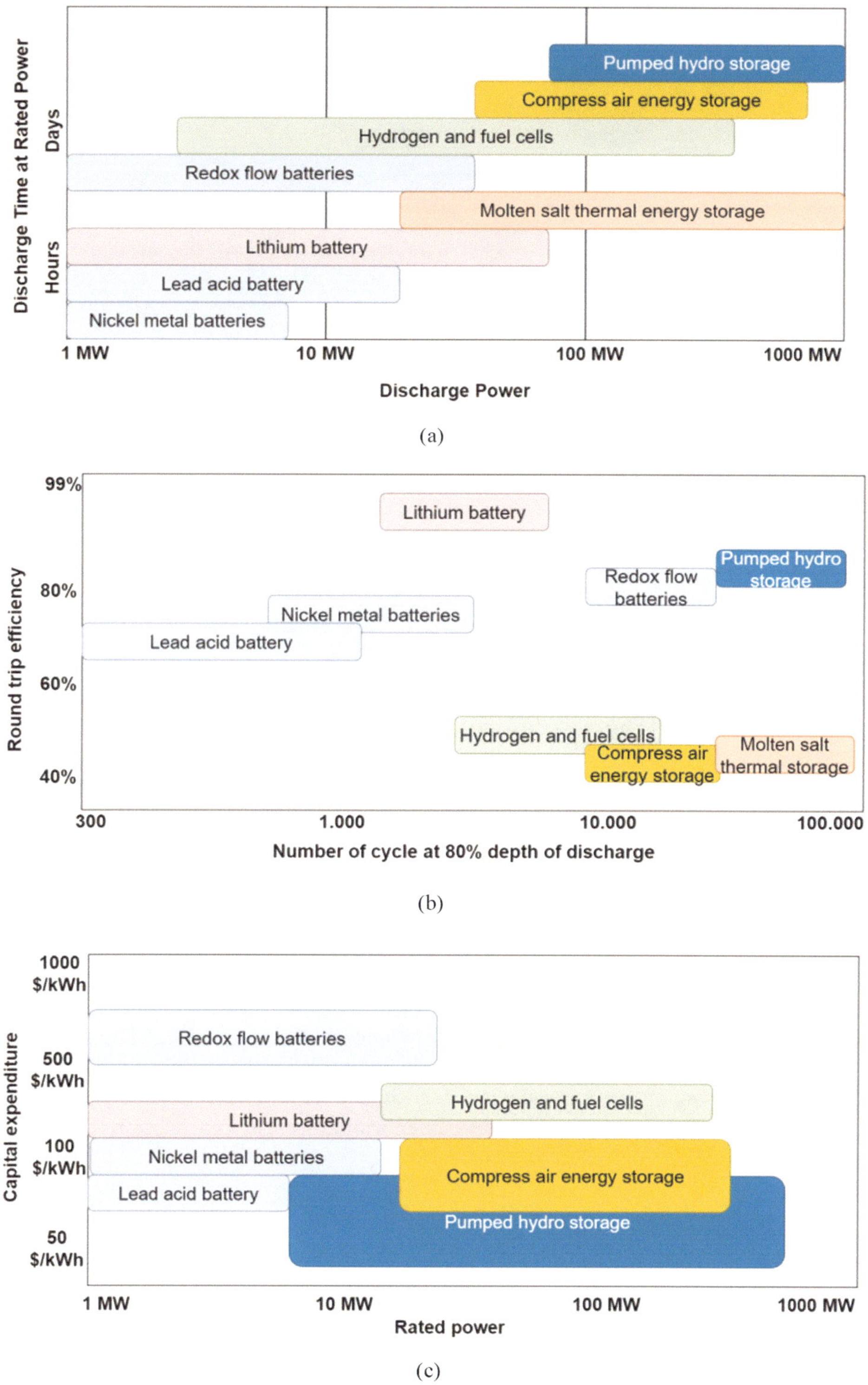

Figure 1.
a: Energy storage technologies discharge time [16]. b: Energy storage technologies efficiency [4]. c: Energy storage technologies capital cost [17, 18].

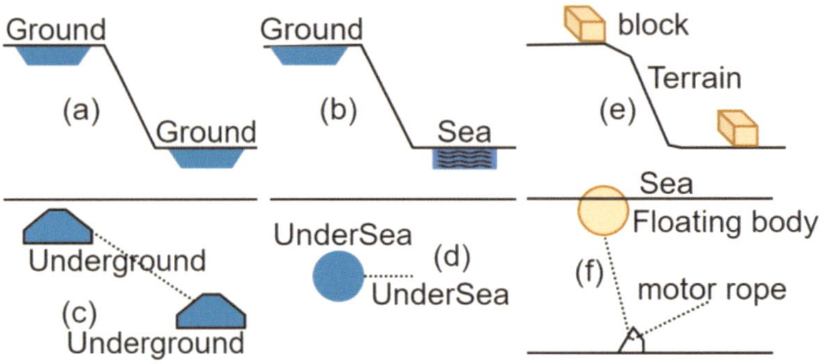

Figure 2.
Gravity energy storage systems topologies.

designs have been proposed by both academics and industries. **Figure 2** summarizes few topologies of GES systems as follow:

3.1 Water based systems topologies

3.1.1 Surface reservoir- surface reservoir

The standard GES is a PHS using a pair of upper-lower reservoirs, both on the ground surface. Existing hydro power plants can be converted into PHS by adding the pumping equipment. The reservoirs are done with a dam wall or from natural depressions. Globally, from the 77 existing PHS plants and the 64 plants under construction, the majority uses dam wall which can range from 10 m up to 150 m in height [16].

3.1.2 Surface reservoir -sea reservoir

To avoid conflict over land usage, and to be close to load centers, using an upper ground surface reservoir and the sea as the lower reservoir has been proposed since 1999 in Japan. However, finding a seaside cliff with a suitable head, near load centers may not be easy globally. In addition, this solution faces the challenges of salt corrosion for the equipment. In Australia, a study concluded that installing a desalination plant to provide fresh water to the system is much cheaper than using corrosion-proof materials or running active corrosion protection equipment [17]. Several projects in Greece, Ireland, Chile, and the USA have been scheduled for years, but not implemented because the funding issues.

3.1.3 Sea reservoir -sea reservoir

This configuration avoids land use by building a spherical storage cavity in the seabed as a lower reservoir and using the sea as an upper reservoir. Studies estimate that such configuration is cost-effective at a deep ranging from 200 to 800 m [18, 20, 21]. Due to the massive weight of concrete used for the sphere (over 2 m thickness, over 30 m diameter), installation is as challenging as the size of the system! Small-scale prototypes have been tested in Europe and USA, however, so far, no commercial proof of concept has been realized.

3.1.4 Underground reservoir -underground reservoir

Underground PHS is attractive as it can be built anywhere, thus eliminating the need of special site or transmission cable. Digging is the most challenging activity of this method. However, sectors such as mining or road tunneling may contribute to a cost-effective implementation of this configuration. The cost drops significantly if underground natural or man-made cavities already exist. For example, some authors proposed the use of decommissioned mines chambers as reservoirs [22, 23]. Also, a single underground reservoir can be connected to a ground surface reservoir made of wall.

3.2 Solid-gas based systems topologies.

3.2.1 Buoyancy energy storage

Buoyancy energy storage (BEST) is another form of GES. By forcing a low-density object (eg. a pressurized air balloon) into the water, the uplift buoyancy force is equal to the weight of the displaced water. This principle is used as the driving force of BEST. A rope attached to a generator allows the pull-down and the uplift of the object for a charge–discharge operation. Studies recommended an anchoring deep of over 3000 m to obtain an energy cost in the range of 50–100 $/kWh [24]. The challenge with BEST is the anchoring stiffness requirements to hold such astonishing force, and the slow speed of the object (<0.1 m/s) needed to minimize the drag losses, meaning the charge–discharge power is also limited. A BEST prototype of 25 m3 balloon at 2 m deep has been tested so far. However, the system suffers from low efficiency due to its small scale [25]. Therefore, more R&D is still needed to get the system to a commercial level.

3.2.2 Solid block gravity energy storage

Instead of using water for power transfer, heavy solid blocks have been proposed as potential energy carrier. As earth or concrete density is nearly 3 times more than water, solid blocks offer more energy density compared to water. In the USA, a company started the testing of an advanced rail energy storage system (ARES) that uses special rails to move tons of blocks uphill to store energy [26]. The ARES can deliver 10 MW of power in 3 seconds while driving downhill. According to company notes, the power cost is expected to be 1200 $/kW compared to 600–8000 $/kW for PHS. Another company conceptualized the use of a large cylinder drilled out of the earth to be used as a hydraulic piston head to store energy in the form of pressurized water [27, 28]. The cost is projected to be 120–380 $/kWh. There are plenty of GES proposals such as using a crane to stack-unstack blocks of concrete in Lego style or using skyscrapers and their elevators to lift-descend blocks. If these methods work in theory, in practice they face extremely challenging factors such as structural cost, wind resistance, limited storage capacity, safety, and more.

4. Megawatt hours scale pumped hydro storage site detection

Due to their relatively small size, modular megawatt hour scale closed loop PHS can be built near communities. Therefore, their site requirements in terms of footprint, head, safety and so on, are much less constraining than those of gigawatt hour scale PHS. In case of a rupture of a reservoir for example, only few hundred thousand cubic meter of water will spill without causing much damage. Given the small footprint of the system, the major requirement of the site is only the height difference as minor excavation or wall construction is sufficient to build the reservoirs. The potential energy and power of water are given in Eq. (1), Eq. (2) as function of water volume, head, flow rate and system efficiency. By assuming a lossless conversion of potential energy from the upper reservoir to kinetic energy at the lower reservoir, the maximum velocity can be expressed as in Eq. (3). The power equation can then be rearranged in Eq. (4), Eq. (5) as a function of water flow diameter, head, and system efficiency. The water specific energy at different heads, as well as the power at different heads and flow diameters are given in **Figure 3**. By assuming a reservoir depth of 10 m, the storage capacity varies from 20 MWh per hectare at 100 m head to 65 MWh per hectare at 300 m head. Therefore, a 10 hectares reservoir can provide up to 650 MWh storage using 1 million cubic meters of water. Regarding the power, for 100 m head, it varies from 2.45 MW at 0.3 m flow diameter to 22 MW at 0.9 m flow diameter. Despite the relatively small flow diameter, these power levels are above the minimum required for 10 hours storage. By increasing the head and the flow diameter, a significantly high power comparable to that of lithium battery can be generated.

$$E(V, \Delta h, \eta) = \rho \, \mathrm{V} \, \mathrm{g} \Delta h \, \eta \tag{1}$$

$$P = \rho \, \dot{\mathrm{V}} \, \mathrm{g} \Delta h \, \eta \tag{2}$$

$$v_{\max} = \sqrt{2 \, \mathrm{g} \, \Delta h} \tag{3}$$

$$P = \rho \, \sqrt{2 \, \mathrm{g} \, \Delta h} \frac{\pi D^2}{4} \, \mathrm{g} \Delta h \, \eta \tag{4}$$

$$P(D, \Delta h, \eta) = \rho \, \sqrt{2} \, \mathrm{g}^{1.5} \frac{\pi D^2}{4} \, \Delta h^{1.5} \, \eta \tag{5}$$

$$E : \text{Energy in joule}$$

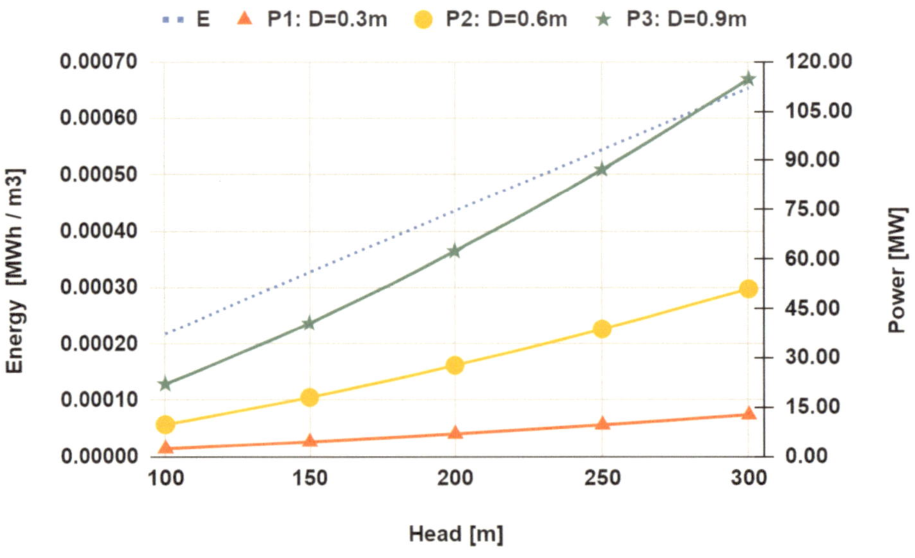

Figure 3.
Water energy and power graph.

$$P : Power\ in\ watt$$

$$v_{\max} : a\ body\ free\ fall\ maximum\ velocity\ in\ meter\ per\ second$$

$$V : water\ volume\ in\ m^3$$

$$\dot{V} : water\ flow\ rate\ in\ cubic\ meter\ per\ second$$

$$\Delta h : reservoirs\ elevation\ difference\ or\ head\ in\ meter$$

$$\eta : Energy\ conversion\ efficency\ in\ percentage$$

$$\rho : water\ density\ in\ kilogram\ per\ cubic\ meter$$

$$g : earth\ gravity\ in\ meter\ per\ second\ square$$

$$D : water\ flow\ diameter\ in\ meter$$

4.1 Site detection methodology

Pumped hydro storage site detection methods follow the same fundamental flow processes [29, 30]. First, geographical information system (GIS) is used to collect digital elevation model and region-specific characteristics such as river, protected areas, ocean and so on. In a second step, GIS data are processed to identify reservoirs. As mentioned previously, there are various topologies of interest such as a pair of dry gully reservoirs, a dam and a river, a dam and the ocean, a dam and an underground reservoir and more [31]. For small scale closed loop PHS, this study is looking for inland, above ground reservoirs to avoid the cost of dealing with corrosive sea water and underground construction. In a third step, the reservoirs are matched according to user criteria such as head, separation distance, and energy capacity. More complete studies add a last step to optimize the reservoirs combination by further taking into consideration the development cost. The illustration in **Figure 4** shows an example of

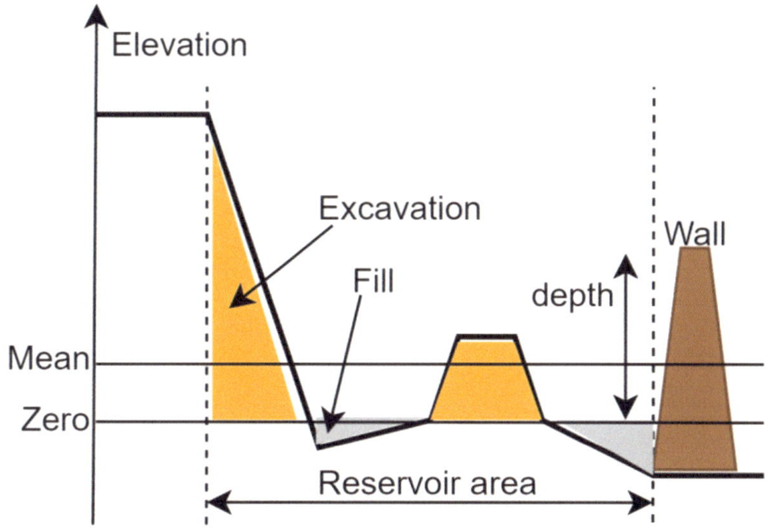

Figure 4.
Reservoir terrain illustration.

reservoir terrain. To build the reservoir, the earth above the marked zero elevation is moved to fill the depression below the zero elevation, and the remaining used to build the wall. On a flat terrain, the excavation work might be few dozen centimeters below the surface to reach the zero level of the reservoir. As the reservoir area increases, less and less excavation is needed to build a wall of certain thickness. Stated differently, to get a fixed volume (wall) the larger the surface (reservoir area), the smaller the depth need to be (excavation).

The methodology summarized in **Figure 5** describes how a pair of reservoirs are detected.

In steps (1)–(2), using global human settlement layer, a community center with population between 100.000 to 500.000 is chosen, then the search zone is delimited by setting a 50 km by 50 km perimeter around the center [32]. This is done to constrain the storage site close to the community. For megawatt hour scale PHS, limiting the eventual connection power line length is important to keep the project unit cost low. A summary of the chosen cities' population, energy consumption per capita and required storage draft estimation is provided in **Table 1**. The storage capacity is calculated by multiplying the daily energy usage by the number of people. The actual number might be lower as some energy may be consumed immediately after production without going through storage first. There is a huge gap in energy consumption between low-income city (Banfora) and high-income city (Syracuse).

In steps (3)–(4), digital elevation model data is pulled from online, a median filter is applied to eliminate eventual outliers, then no go zones are marked [33–35]. If the algorithm find a reservoir that overlap partially or entirely with a marked zone, the solution is discarded. For simplicity, only the region around the city center with visible housing concentration has been excluded.

In step (5) a mask for both reservoirs which are assumed to be 10 m height is created. The elevation standard deviation is set to maximum 5 m to limit the civil

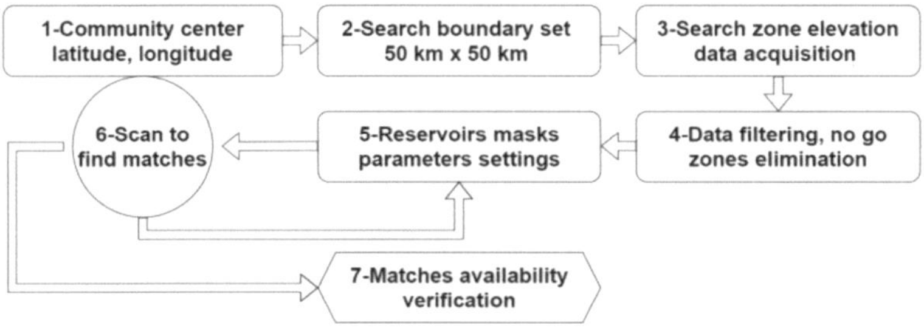

Figure 5.
Pumped hydro storage reservoirs pairs detection method.

City	Population	kW/Capita/year	MWh needed
Banfora	122.000	100	33
Syracuse	253.000	12.000	8.318
Manisa	294.000	3.000	2.416
World	8.000.000.000	3.000	65.753.425

Table 1.
Selected cities data.

engineering work required to build the reservoirs. The surface parameters are set to a square of 1 ha, 4 ha, and 9 ha while the head parameter is set to a minimum of 100 m. Lastly, the distance between two reservoirs is set to a maximum 1000 m. By reference, giga scale PHS reservoirs can have a separation distance tens of kilometers, making the construction process more difficult and costly.

In step (6) iteration is done to find pairs of reservoirs that satisfy each set of parameters described in (5). From an elevation point, the average surface elevation corresponding to the first reservoir is calculated, then within 1000 m from that point, a surface with average elevation that satisfy the head requirement is kept as a second reservoir. The excavation and fill volumes are calculated as they can serve to rank the attractiveness of the sites, the less the civil work the best. The water volume is 100.000 m3/ha corresponding to 20 MWh per 100 m head at 75% round trip efficiency. This leads to a storage capacity of 20 MWh at least to 540 MWh if the largest surface is matched with 300 m head. If the head reach 500 m the maximum capacity increases to 900 MWh which is still in the sub gigawatt hour scale.

In step (7), candidate sites are displayed in a map. A local inspection which is out of scope of the study must be carried out to confirm the availability of the site. Also, the type of the soil must be sampled to assess the technical and economical attractiveness of the land for project development. The aims of the paper is to simply to orient researchers and industry attention toward potentially feasible storage option that is under looked by standard search criteria.

4.2 Site detection results

The results of megawatt-hours scale neighborhood closed loop PHS sites detection is shown in **Table 2** for 3 cities, Banfora, Syracuse and Manisa. When the reservoirs areas are set to 1 ha, there are 556, 711 and 1966 pairs detected for the above cities respectively. For Manisa, the number is exceptionally high as the city is surrounded by mountains and depressions. For Banfora and Syracuse, when each reservoir area is increased to 4 ha or 9 ha, the number of pairs reduce significantly by a factor of 10 approximately. For Manisa, the reduction is more pronounced. In general, the more uneven a terrain is, the more challenging it is to find larger reservoir with the 5 m standard deviation set to reduce civil engineering work. Manisa terrain is more accidented than Banfora and Syracuse which explain the drastic drop in the number of reservoirs with 4 ha or 9 ha area. However, it is worth noting that the reduction in pair count does not necessarily reduce significantly the total storage capacity. For Banfora and Syracuse, 9 ha reservoirs capture over 87% of 1 ha capacity. This means that modular storage projects (1 ha) can be done on areas with higher potential (9 ha) to save cost on future expansion by sharing existing infrastructures (roads, power lines). The available storage capacities are 39,300%, 192%, and 1945% of the required

	Banfora (Burkina)			Syracuse (New York)			Manisa (Turkey)		
Area [ha]	**n [−]**	**H [m]**	**E [MWh]**	**n [−]**	**H [m]**	**E [MWh]**	**n [−]**	**H [m]**	**E [MWh]**
1 ha	556	118	13,000	711	116	16,000	1966	119	47,000
4 ha	85	124	8400	93	118	8800	109	124	10,000
9 ha	54	127	12,000	65	120	14,000	11	123	2400

Table 2.
Sites detection results.

capacities for Banfora, Syracuse, and Manisa respectively. Even for 9 ha reservoirs, in all cases, the storage potential is enough to meet or exceed the need of the cities, giving the opportunity to power system planners to choose the reservoirs configurations that fits the best to local economic, social, and environmental requirements.

A MATLAB plot of pairs of reservoirs as well as their display on a map are given in **Figures 6** and **7**. The slope of the head can be visualized to assist in practical sites selection. On the plots, the upper and lower reservoirs are visible in color scale. Both reservoirs are on a nearly flat terrain, indicating the algorithm performance in detecting sites with minimal elevation standard deviation. The sites tend to be away from the cliff, thus avoiding unstable and expensive construction. As each grid cell is approximately 30 m by 30 m, the proximity of the pairs can be seen from the plots.

The map shows that for Banfora and Syracuse, the pairs are clustered in four or five regions, while for Manisa they are more dispersed. With clustered sites, more pairs can be combined to create larger storage capacity if needed. Also, infrastructure such as power transmission cable, roads and water pipes can be shared among pairs. With distributed sites, a more resilient energy system can be achieved by connecting each pair to a group of loads in the cities.

The PHS heads distribution is shown in **Figure 8**. The majority of sites have heads between 100 m and 140 m approximately. **Table 2** above shows a mean head around 120 m for all sites. As the cut off head was set to 100 m, the algorithm registers sites as soon as the requirement is met and mark the land unavailable for further usage. This can eliminate the possibility of having higher head with different reservoirs combination. To avoid this missed opportunity, the search started with head set to 400 m, 300 m and 200 m to isolate eventual best sites. The results revealed the existence of few pairs with high head, so the whole study was done with 100 m cut off head. The minor sites with head between 150 m and 230 m are still detected as shown in **Figure 8**. For project implementation it is important to make an extensive parametric study to detect the optimum sites by taking into account not only the head, but also other factors such as the ground characteristics, power line availability, access roads, initial water and top up water collection system, and the acceptance of project by locals.

5. Benefit of multi-disciplinary problem solving

Modern energy system implementation involves many people from various backgrounds. With the increasing complexity of the system, it is crucial to coordinate in a smart way, the integration of ideas and solutions to avoid future problems generation or a solution being outdated before the project end life. Therefore, both engineering and social science aspects of the system must go through a deep analysis and experiments before large scale field applications.

5.1 Engineering

As simple as it seems, gravity energy storage system require knowledge from different engineering fields. For the specific case of PHS, traditionally, civil engineering for excavation and dam construction has been the main work. However, in a closed loop system, other requirements such as water ground infiltration barrier, top up water collection from rain or from underground well may call for new expertise from material science and hydrology fields. Water evaporation also must be minimized for good system performance. Electrically, hydropower systems operations in the last decades were based

Figure 6.
Sample upper and lower reservoirs detected. (top-Banfora, middle-Syracuse, bottom-Manisa).

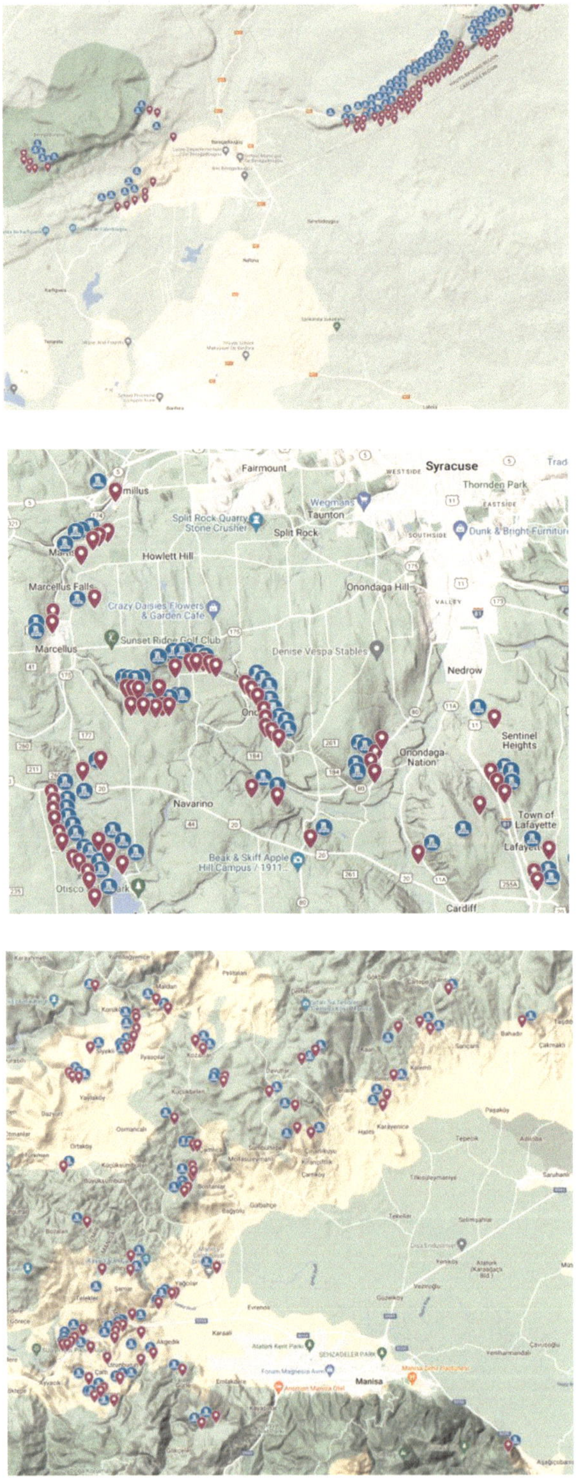

Figure 7.
Mapping of reservoirs pairs. (top-Banfora, middle-Syracuse, bottom-Manisa).

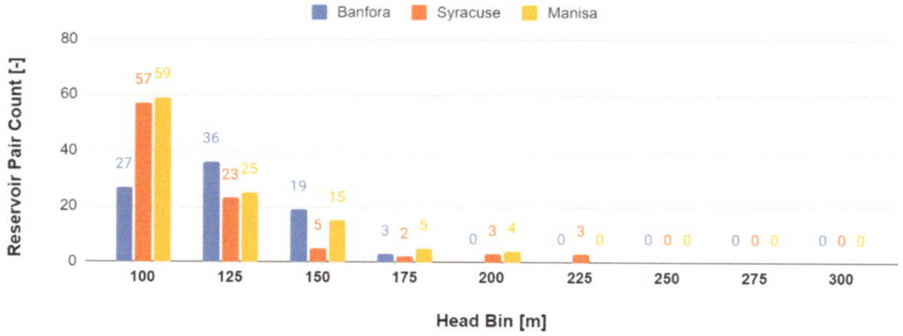

Figure 8.
Sites head distribution.

on a lookup table style optimization which is not ideal when more variable factors such as loads, wind, solar radiation, water quality, system age are taken into account. New fields such as artificial intelligence, machine learning provide better performance for multi-variables optimization and forecasting so they can boost the operation performance of modern PHS. In the case that sea reservoir must be used, decades long experience in offshore oil and ships industries can help to solve corrosion issues. Although multi-disciplinary approach to energy problem is recommendable, caution must be taken to avoid transferring costly methods from other applications to energy storage applications. For this reason, the input of people without a link to the areas mentioned above can be useful for reaching out-of-the-box innovative, cost-effective solutions.

5.2 Regulations

On the social side, the support of both private and public institutions will be beneficial to starting gravity energy storage industry. As the numbers show, energy infrastructures capital expenditures are costly so financial subsidies, research grants, and startups funding will be necessary to experiment with different possibilities before reaching commercial-level solutions. The permitting paperwork constitutes a barrier for some projects, meaning that regulations must be adapted to capture multiple objectives optimized socio-economic values. In the USA for example getting only approval for hydropower project may take 3–5 years, the construction process also 3–5 years, dragging project completion lead time to a decades. Very few private investors can cope with such long-term market uncertainties [36]. A differentiated regulation for giga scale storage project and mega scale projects can be a tool to shorten the development time of closed loop megawatt hours PHS. Also, from country to country, there is a huge gap in energy storage valuation which can be addressed to speed up investments. For example, in the USA, time of use rates and ancillary services market provide distinct value to storage system, while in many other countries, such market operation does not exist. As the world is moving toward solar and wind, power system inertia will be reduced, threatening the system stability. By designing a market for inertia supply, PHS can capture naturally a value without the need of a virtual inertia generation system which is required for storage such as chemical batteries.

6. Conclusion

There are many candidate technologies for large-scale electric energy storage. Up to date, PHS is the dominant technology with over 90% of world storage capacity.

Pumped hydro storage is the future for firming intermittent renewable energy from technology maturity and cost perspective if a suitable site is available. The study revealed that for the selected cities, the storage potential is sufficient to meet the need for 100% renewable energy scenario, even under the assumption that all energy used will go through storage first, without a direct consumption. Banfora with a lower consumption per capita has storage capacity 393 times the current need which means that it can meet future need and become power exporter. Even if the study is limited geographically, previous works from NREL in the USA and IRENA in Australia highlighted that the available global PHS capacity far exceed the need. This study limited the search to neighborhood small PHS sites to serve as a top priority for distributed, resilient energy infrastructure development. This avoids a high transmission cost and environmental disturbance. it is highly recommended to explore distributed megawatt-hours scale closed loop pumped hydro energy storage along with community solar PV or wind projects. The following recommendation can help accelerated PHS investments:

- Provide grants for closed loop megawatt hours scale PHS pilot projects.
- Develop regulatory framework for quick licensing of 20–500 MWh storage projects.
- Design a market that rewards PHS for ancillary services.
- Develop a special feed in tariff to pay for all the values provided by the project, not only limited to energy sales, but also to deferred investment on grid upgrade, peak reduction.
- Develop an algorithm that map investment ready PHS sites (with environmental impact analysis report, construction permit and grid connection authorization).

Author details

Kiswendsida Elias Ouedraogo[1*], Pinar Oguz Ekim[1] and Erhan Demirok[2]

1 Department of Electrical Engineering, Izmir University of Economics, Izmir, Turkiye

2 TESCOM Elektronik A.S., Izmir, Turkiye

*Address all correspondence to: ouedraogo.elias@gmail.com

References

[1] Emrani A, Berrada A, Bakhouya M. Modeling and performance evaluation of the dynamic behavior of gravity energy storage with a wire rope hoisting system. Journal of Energy Storage. 2021;**33**: 102154. DOI: 10.1016/J.EST.2020.102154

[2] Shan R, Reagan J, Castellanos S, Kurtz S, Kittner N. Evaluating emerging long-duration energy storage technologies. Renewable and Sustainable Energy Reviews. 2022;**159**:112240. DOI: 10.1016/J.RSER.2022.112240

[3] Møller KT, Jensen TR, Akiba E, Wen Li H. Hydrogen - A sustainable energy carrier. Progress in Natural Science Materials International. 2017;**27**(1): 34-40. DOI: 10.1016/J.PNSC.2016.12.014

[4] Ibrahim H, Ilinca A, Perron J. Energy storage systems—Characteristics and comparisons. Renewable and Sustainable Energy Reviews. 2008;**12**(5):1221-1250. DOI: 10.1016/J.RSER.2007.01.023

[5] NREL study backs hydrogen for long-duration storage – pv magazine USA. Available from: https://pv-magazine-usa.com/2020/07/03/nrel-study-backs-hydrogen-for-long-duration-storage/ [Accessed: November 18, 2022]

[6] System Operation: Innovation Landscape Briefs. Available from: https://www.irena.org/Publications/2020/Jul/System-Operation-Innovation-Landscape-briefs [Accessed: November 18, 2022]

[7] World Small Hydropower Development Report | UNIDO. [Online]. Available from: https://www.unido.org/our-focus-safeguarding-environment-clean-energy-access-productive-use-renewable-energy-focus-areas-small-hydro-power/world-small-hydropower-development-report [Accessed: November 18, 2022]

[8] Ali S, Stewart RA, Sahin O. Drivers and barriers to the deployment of pumped hydro energy storage applications: Systematic literature review. Clean Engineeirng and Technology. 2021;**5**:100281. DOI: 10.1016/J.CLET.2021.100281

[9] Renewable Energy Cost Analysis - Hydropower. Available from: https://www.irena.org/publications/2012/Jun/Renewable-Energy-Cost-Analysis—Hydropower [Accessed: November 18, 2022]

[10] Construction cost data for electric generators installed in 2020. Available from: https://www.eia.gov/electricity/generatorcosts/ [Accessed: November 18, 2022]

[11] I. R. E. Agency. Renewable Power Generation Costs in 2019, 2020 [Online]. Available from: https://www.irena.org/-/media/Files/IRENA/Agency/Publication/2018/Jan/IRENA_2017_Power_Costs_2018.pdf [Accessed: November 18, 2022]

[12] Görtz J, Aouad M, Wieprecht S, Terheiden K. Assessment of pumped hydropower energy storage potential along rivers and shorelines. Renewable and Sustainable Energy Reviews. 2022; **165**:112027. DOI: 10.1016/J.RSER.2021.112027

[13] Stocks M, Stocks R, Lu B, Cheng C, Blakers A. Global atlas of closed-loop pumped hydro energy storage. Joule. 2021;**5**(1):270-284. DOI: 10.1016/J.JOULE.2020.11.015

[14] Rosenlieb E, Heimiller D, Cohen S. Closed-Loop Pumped Storage Hydropower Resource Assessment for the United States. Final Report on HydroWIRES Project D1: Improving

Hydropower and PSH Representations in Capacity Expansion Models. United States. DOI: 10.2172/1870821

[15] NHA Releases 2018 Pumped Storage Report - National Hydropower Association. Available from: https://www.hydro.org/news/nha-releases-2018-pumped-storage-report/ [Accessed: November 18, 2022]

[16] List of pumped-storage hydroelectric power stations - Wikipedia. Available from: https://en.wikipedia.org/wiki/List_of_pumped-storage_hydroelectric_power_stations [Accessed December 09, 2022]

[17] Ansorena Ruiz R et al. Low-head pumped hydro storage: A review on civil structure designs, legal and environmental aspects to make its realization feasible in seawater. Renewable and Sustainable Energy Reviews. 2022;**160**:112281. DOI: 10.1016/J.RSER.2022.112281

[18] Dick C, Puchta M, Bard J. StEnSea – Results from the pilot test at Lake Constance. Journal of Energy Storage. 2021;**42**:103083. DOI: 10.1016/J.EST.2021.103083

[19] Order Megapack. Available from: https://www.tesla.com/megapack/design [Accessed: December 02, 2022]

[20] Dubbers D. Comparison of underwater with conventional pumped hydro-energy storage systems. Journal of Energy Storage. 2021;**35**:102283. DOI: 10.1016/J.EST.2021.102283

[21] Slocum AH, Fennell GE, Dundar G. MIT open access articles ocean renewable energy storage (ORES) system: Analysis of an undersea energy storage concept terms of use: Creative commons attribution-noncommercial-share alike 3.0. Proceedings of the IEEE; **101**(4):906-924. DOI: 10.1109/JPROC.2013.2242411

[22] Kitsikoudis V et al. Underground pumped-storage hydropower (UPSH) at the martelange mine (Belgium): Underground reservoir hydraulics. Energies. 2020;**13**(14):3512. DOI: 10.3390/EN13143512

[23] Morabito A. Underground cavities in pumped hydro energy storage and other alternate solutions. In: Encyclopedia of Energy Storage, Spain. 2022. pp. 193-204. DOI: 10.1016/B978-0-12-819723-3.00145-1

[24] Hunt JD et al. Buoyancy energy storage technology: An energy storage solution for islands, coastal regions, offshore wind power and hydrogen compression. Journal of Energy Storage. 2021;**40**:102746. DOI: 10.1016/J.EST.2021.102746

[25] Bassett K, Carriveau R, Ting DSK. Underwater energy storage through application of Archimedes principle. Journal of Energy Storage. 2016;**8**:185-192. DOI: 10.1016/J.EST.2016.07.005

[26] Cava F, Kelly J, Peitzke W, Brown M, Sullivan S. Advanced rail energy storage: Green energy storage for green energy. In: Storing Energy with Special Reference to Renewable Energy Sources. South Africa. Elsevier; 2016. pp. 69-86. DOI: 10.1016/B978-0-12-803440-8.00004-X

[27] Botha CD, Kamper MJ. Capability study of dry gravity energy storage. Journal of Energy Storage. 2019;**23**:159-174. DOI: 10.1016/J.EST.2019.03.015

[28] Gravity Storage - a new technology for large scale energy storage. Available from: https://heindl-energy.com/ [Accessed: December 2, 2022]

[29] Ouchani F, Jbaihi O, Alami Merrouni A, Ghennioui A, Maaroufi M. Geographic information system-based multi-criteria decision-making analysis for assessing prospective locations of pumped hydro energy storage plants in Morocco: Towards efficient management of variable renewables. Journal of Energy Storage. 2022;**55**:105751. DOI: 10.1016/J.EST.2022.105751

[30] Rogeau A, Girard R, Kariniotakis G. A generic GIS-based method for small pumped hydro energy storage (PHES) potential evaluation at large scale. Applied Energy. 2017;**197**:241-253. DOI: 10.1016/J.APENERGY.2017.03.103

[31] Lu B, Stocks M, Blakers A, Anderson K. Geographic information system algorithms to locate prospective sites for pumped hydro energy storage. Applied Energy. 2018;**222**:300-312. DOI: 10.1016/J.APENERGY.2018.03.177

[32] Global Human Settlement - GHSL Homepage - European Commission. Available from: https://ghsl.jrc.ec.europa.eu/ [Accessed: November 18, 2022]

[33] Elevation API |Available from: Open-Meteo.com. https://open-meteo.com/en/docs/elevation-api [Accessed: November 18, 2022]

[34] EarthExplorer. Available from: https://earthexplorer.usgs.gov/ [Accessed: November 18, 2022]

[35] Overview | Elevation API | Google Developers. Available from: https://developers.google.com/maps/documentation/elevation/overview [Accessed: November 18, 2022]

[36] Pumped Storage Report 2018. Available from: https://www.hydro.org/wp-content/uploads/2018/04/2018-NHA-Pumped-Storage-Report.pdf [Accessed: December 02, 2022]